LAWS ON
NUCLEAR ISSUES

Encyclopaedia of Environmental Pollution and Awareness in 21st Century Series

LAWS ON NUCLEAR ISSUES

Edited by

V.K. Prabhakar

ANMOL PUBLICATIONS PVT. LTD.

NEW DELHI - 110 002 (INDIA)

ANMOL PUBLICATIONS PVT. LTD.
4374/4B, Ansari Road, Daryaganj
New Delhi - 110 002

Laws on Nuclear Issues

First Edition, 2001

ISBN 81-261-0913-0

PRINTED IN INDIA

Published by J.L. Kumar for Anmol Publications Pvt. Ltd., New Delhi - 110 002 and Printed at Mehra Offset Press, Delhi.

CONTENTS

PREFACE

Today, the cry of 'pollution' is heard from all the nooks and corners of the world, and pollution has become a major threat to the very existence of mankind on this earth. It is the major challenge of our times.

The pollution of various resources has gone to such an extent that we are unable to breathe fresh air and drink fresh water. On one hand the advancements of science and technology have added to the human comforts by giving us automobiles, electrical appliances, supersonic jets, space crafts, better medicines, better chemicals to control harmful insects and other pests, etc., but, on the other hand, they have given us a very serious problem to face that is pollution. Pollution can be defined as an undesirable change in the physical, chemical, or biological characteristics of the air, water, or land that can harmfully affect health, survival, or activities of humans or other living organisms.

Pollution does not have to cause physical harm, but it may merely interfere with human activities. Chemicals spewed into the air or water from an industrial plant may be harmful to humans and other organisms living near the plant. If expensive pollution controls are required, the plant may be forced to shut down.

India faces serious problems of environmental pollution. Water pollution from untreated community and industrial wastes flowing into our rivers and streams, increasing use of fertilizers and pesticides, dumping of organic wastes etc., are issues that need to be tackled on the priority. Air pollution, long regarded as the bane of industrialised nations looms large over our country.

Here, we are presenting this *Encyclopaedia of Environmental Pollution and Awareness in Twenty-first Century* in fifty volumes with a view to make people conscious about the seriousness of pollution in the present time and its remedies.

Editor

1 NUCLEAR POLLUTION

The far-reaching portion of the atmosphere, called the magnetosphere, is having an area of high-energy radiation called the Van Allen region. It is roughly doughnut shaped, extending from about 500 miles above the earth at the equator to an altitude of about 40,000 miles. The Van Allen region was at first thought to present a major obstacle to space travel, but further studies revealed that astronauts would pass through the zones of high radiation quickly and that precautions could be taken to disallow excessive exposure.

We have now learned how to produce radiation in many forms, and our exposure to the normal level of environmental radiation that remained relatively constant for millions of years has been greatly increased. We are being bombarded with man-made radiation from x-ray machines, radioactive fallout from nuclear explosions, effluent from nuclear power plants, radioactive materials and other radiation sources in research laboratories, industrial plants and hospitals, wastes from reprocessing nuclear fuels, and from mining and processing radioactive products. Nuclear pollution is caused due to the addition, through activities of man, of ionizing radiations to the environment giving people an exposure to more of such radiations than they normally would experience. Radiation is able to permeate the universe, the solar system, and the earth.

Extremely small amounts of radiation can be profoundly injurious, although different forms of radiation are having different biological effects. Visible light and infra-red heat rays are kinds of radiation that are generally beneficial, but some of the more potent forms of radiation, for example, x-rays, destroy parts of living cells and tissues.

As these highly energetic forms of radiation tend to split substances, including matter, into ions, they are known as ionizing radiation. The various forms of ionizing radiation differ greatly in their penetrating power and in the intensity of their injurious effects. Sun has been the source of spectrum of radiations like radiowaves, infra-red, ultraviolet, x-rays, gama and cosmic rays.

In addition, radioactive isotopes give off subatomic particles like protons, neutrons, electrons and helium nuclei called x-particles in the process of decomposition from an unstable state to a stable state to a more stable condition. Radiation is a form of energy which may be transferred from one body to other through empty space. Radiation has been of two main types. Sunlight—the form of radiation people are most familiar with—is an example of the type called electromagnetic radiation. It covers a broad spectrum of radiant energy and a wide range of physical and biological effects. The other type, called particulate radiation, consists of parts of atoms which are travailing at high speeds, often with tremendous energy.

Both electromagnetic and particulate radiation are having great biological importance and account for some of the most serious environmental pollution problems. Electromagnetic radiation consists of a broad spectrum of energy having the nature of light. Radiowaves, infra-red rays, visible light rays, ultraviolet rays, x-rays, and gamma rays have been all part of the electromagnetic spectrum. All the different kinds of electromagnetic radiation have been nothing more than light rays of different wavelength and frequency. The fact that our eyes have been sensitive only to visible light, actually a very small part of the light spectrum, has been enough to remind us how different the world would look if, through some accident of nature, we could also see x-rays or radio-waves.

The energy of electromagnetic radiation gets transmitted in packets called photons. All photons, of whatever part of the spectrum, travel at a speed of 186,282 miles per second, the speed of light. During their flight though space, they seem to vibrate, so that the rays have a wavelike quality. The energy of a photon is considered as being inversely proportional to its wavelength. It means that the shorter the wavelength, the more energy there is in the photon. Therefore, as the wavelength of the electromagnetic spectrum decreases, the radiation has been said to become more energetic.

For example, the visible rays of sunlight have been more energetic than radio rays of longer wavelength, and ultraviolet light is more energetic than visible light. X-ray having a wavelength of 0.1 nanometer (1 A^0) possess 3,000 times as much energy as ultraviolet rays having a wavelength of 300 nanometers. As the wavelength decreases and the energy of the radiation increases, the potentialities to disrupt living tissue would become very great.

The electromagnetic wave has two component parts —an electric and a magnetic moving in phase, but in direction 90^0 from each other much like two vibrating strings—one going up and down and the other back and forth, superimposed on each other. The wave carries no net electrical charge and no net magnetic moment but because of the components which can interfere or react with electric or magnetic fields, it can lose or gain energy (i.e., change frequency).

These are radiowaves in or near the extremely high frequency, or shorter wavelength, range. Hence, they are electromagnetic radiations in the more energetic portion of the radio wave spectrum. Microwave energy has been much too low to disrupt living tissues by ionization. Instead, the energy gets absorbed as oscillation energy and is converted to heat. This makes it possible to use microwaves for cooking. But the same radiations have been capable of heat injury to tissues in the body when exposed to the radiations, thus making it necessary to carefully maintain the door seals of microwave ovens.

While microwaves can bring about bodily damage with heat, most of our discussion will be dealt with the more energetic radiations that can disrupt tissues, including cellular structures, by ionization, or splitting, of molecules. Ultraviolet light is electromagnetic radiation just beyond the shortest wavelength of violet light to which the eye has been sensitive.

It is having profound physical and biological effects. Though shorter in wavelength than the most energetic rays of visible light, ultraviolet light rays are longer than the average for X-rays overlapping the shorter and more energetic x-rays. Ultraviolet light covers from 2-180 nm in vacuum and 180-400 nm for far and near UV of electronagnetic spectra. Ultraviolet light is produced in abundance by the sun and can be generated artificially by the carbon arc or the mercury vapor arc. Ultraviolet radiation penetrates human skin only superficially, probably to a maximum depth of 1 millimeter, which, however small is enough to cause serious injury. Ordinary glass about 2 millimeters thick filters out (absorbs) ultraviolet rays shorter than 310 nanometers. It has been fortunate that the air, smoke, and dust filter out much of the ultraviolet radiation from the sun, especially the shorter wavelengths, for otherwise the damaging effects would drastically change life on earth. The shorter wavelength ultraviolet radiation near the x-ray region can bring about damaging ionization.

These are produced when a fast moving electron is suddenly accelerated/decelerated, releasing electro-magnetic (em) radiations called bremsstrahlung. Another mode of production has already been discussed before. X-rays include a broad spectrum of wavelength. Therefore, the energy, penetrating power, and disruptive force of x-rays differ greatly depending on the wavelength. An "average" x-ray is having a wavelength of approximately 0.1 nanometer, a distance about equal to the diameter of a molecule of water.

The discovery of x-rays was done by German physicist Wilhelm K. Roentgen in 1895. He was working on the luminescence of chemicals produced by cathode rays when he noticed a strange phenomenon occurring some distance from the tube. A sheet of paper coated with a luminescent substance (barium platinocyanide) was glowing. As the cathode tube was blocked off by cardboard, it was not at first apparent that the rays from the cathode tube could cause the luminescence. Roentgen reasoned that there would be some kind of penetrating radiation that was invisible to the eye.

He experimented and reported that the radiation could penetrate thin layers of metal. Because he had no idea of what the strange radiations could be, he called them by the traditional symbol for the unknown, "x" rays. The name continued to be used after the nature of x-rays became known. For his excellent work, Roentgen received the Nobel Prize in 1901, the first year of its award.

Gamma rays are similar to x-rays, with which their wavelengths overlap, but because most gamma rays are of shorter wavelength, they generally have higher energy and are more penetrating than x-rays. Short wavelength x rays, sometimes called hard x-rays, may be identical to gamma rays at the longer wavelength end of their spectrum. Gamma rays are having wavelengths of about 0.1 nanometer and shorter.

Regardless of the wavelength, the term gamma rays is used only in reference to electromagnetic radiation that is emitted from the atomic nucleus of a radioactive element. For example, some atoms of radium, uranium, and plutonium give off gamma rays. These are produced by change in the energy levels of nucleus of atoms. Wavelength of this radiation varies from 0.0003 to 0.1 nm. The common sources are cobalt-60 and cesium,-137. The radioisotope of cobalt has been the source of two monochromatic gamma beams with energies of the gamma quanta 1.17 and 1.33 MeV. The half-life is 5.3 years. Cobalt-60 has been the most widely used source of gamma rays for radiochemical and radiobiological studies.

Cesium-137 is a radioisotope having a half-life of 30 years and gives both electrons and gamma radiations. Gamma ray energy is about 0.662 MeV. It is also widely used as a source of radiation for therapy, radiobiological studies etc. Other sources of gamma rays have been (a) the spent fuel rods from nuclear reactors, which are actually blocks of uranium, and (b) products of fission.

Particulate radiation gets produced when one or more of the various components of the atom, for example, an electron, a proton, or a neutron, have been emitted from a radioactive element. Such particles may get ejected from the atom with great force, sometimes along with gamma rays as the atom gives up its energy, or in technical terms, as it gets decayed to a less energetic form of matter. The time it takes one-half a the nuclei of the atoms in a sample of radioactive element to decay is termed as the half-life. But whether the radiation from nuclear disintegration is electromagnetic or particulate, the emanations have been so energetic and forceful that they can do great damage to living cells and tissues.

Beta radiation is made up of high-speed electrons called beta particles. They get emitted by many of the radioactive substances during their decay to more stable states. An electron is exceedingly small; it weighs only 1/ 1,835 as much as a proton. When an electron is rejected at high speed from the atom of a radioactive element, it ionizes substances it collides with in its path.

Thus beta particles damage tissue. But they are less penetrating than x-rays or gamma rays. Carbon 24, phosphorus 32, hydrogen-3 (tritium), and strontium-90 are examples of beta emitters. Electrons with energy range 3 to 100 MeV can be obtained from the various types of accelerators, such as the Van-de-Graaff generator. Such installations can produce electrons with energies of upto 10^2 MeV and dose rates of the order of 10^{10} roentgen (R) per hour. Aside from these generators, there are many other radiation sources such as betatrons, linear accelerators, etc.

Some radioactive isotopes emit fast moving particles, called alpha particles, containing 2 protons and 2 neutrons. This is equivalent in make-up to a helium nucleus. Since alpha particles contain no electrons, they are positively charged. Alpha particles are less penetrating than either beta particles or x-rays and gamma rays. They could be stopped by a sheet of paper. Alpha radiation, therefore, is dangerous only when the emitter is in direct contact with tissue, such as when inhaled or ingested. But in such cases the alpha particles are readily absorbed by tissue, so the danger is very great. Radium-226, uranium-238, and plutonium-239 are examples of radionuclides that emit alpha particles.

Protons are 1,835 times heavier than electrons. While the energetic electron (beta particle) drives into tissue like a tiny particle of sand, the proton lumbers along like a rock, knocking off pieces of atoms and molecules at it goes. The proton does not penetrate as far as an electron of the same energy, but it causes more disruption in a small area. It has been said to leave a heavier track. The effect has been similar to that of an alpha particle.

This is a form of beta-decay. The practical importance of neutrons comes from the fact that a neutron can stimulate a nucleus of a fissile substance such as plutonium to undergo fission, with a release of energy. Neutrons can also be absorbed by 'fertile' nuclei (e.g. thorium) to produce additional fissile material. Neutrons will also be absorbed by the metal lithium with the production of the fusion fuel tritium.

In fission reactors, neutrons are both a product and a cause of fission, a fact underlying the chain reaction concept. However, to ensure proper reactor operation, the flux of neutrons has to be carefully controlled-by moderators which slow neutrons down, by absorbers which remove them from circulation and by reflectors with the effect of preventing neutron loss by turning them back into the reactor core. Neutrons are copiously given off by the cores of fission reactors and can be a radioactive hazard through neutron activation; on the other hand their application in fuel-breeding makes neutrons very precious in the sense that they can be used to multiply man's energy resources many times over through the use of thorium and the heavy form of uranium (^{238}U).

Neutrons are having approximately the same mass as protons but have no charge. They are obtained from either a neutron beam or from the atomic disintegration of one of the radioactive isotopes, for example, uranium-235. Fast neutrons have been four or five times more lethal to mice than x-radiations. Neutrons shot from uranium-235 was the basis for the chain reaction of uranium in Fermi's famous "pile" under the bleachers of the stadium at the University of Chicago in 1942, where experiments on the first atomic bomb were conducted. Major sources of neutrons have been nuclear reactors for irradiation of chemical and biological specimens.

These give a mixed emission, composed of a stream of neutrons (thermal and fast) and gamma rays. The intensity of the stream of thermal neutrons has been of the order of10^5 to 10^9 R/hour. Neutrons are also available from neutron generators and cyclotrons in which low Z nuclei (H^2, H^3, Hc^4,Be^9) are bombarded by high velocity and positively charged particles. In order to get beams of heavy particles of very high energy, the various types of accelerators are employed. For example, the Van-de-Graaff accelerators are able to produce protons of an energy level ranging from 1 to 10 MeV.

Small cyclotrons provide beams of accelerated protons, deuterons and alphaparticles with energies of upto 10 MeV and with a penetrating ability (in biological tissue) of the order of a fraction of a millimeter. Beams of greater penetrating power (with energies of several tens of MeV) are having path length of the order of several centimeters.

In order to get beams of even higher intensity, linear accelerators can be used in series with a cyclotron. In that case, the object of irradiation must be properly cooled so as to disallow the over heating, and consequently undesirable side reactions. Cosmic rays are energetic subatomic particles, from the sun and outer space. They strike the earth at high velocities, some of them penetrating several thousand feet of solid rock. Cosmic rays are part of the background radiation that affects all living organisms. The dosage from cosmic rays at sea level is estimated to be equal to about 0.001 roentgen per day. The amount of radiation that can be tolerated by humans without detectable injury is about 0.01 roentgen per day. Because the difference between background cosmic radiation and detectable human tolerance is small (about ten-fold), cosmic rays probably have significant effects on cells and tissue, no doubt causing mutations and other chromosomal aberrations. It has been estimated that cosmic rays account for about one-fourth of the mutations caused by radiation.

In its broadest sense, radiation is energy being propagated from one place to another through space. There are maiuly two types of radiations:

1. Non-ionising radiations
2. Ionising radiations

Ionizing radiations are the potent sources of energy. Their toxicity is almost 100 million times more than that of cyanides and other reagents.

Non-ionising radiations

Radiations of shorter wavelength but having greater energy may be able to harm the microorganisms but are able to injure only the surface tissues of higher plants and animals. These radiations are also known to increase the rate of mutations.

Some plants are able to absorb maximum radiation in the ultraviolet region, they can grow well in green houses where practically all the ultraviolet light has been filtered out. This shows that ultraviolet light is not required for plant growth. In biological systems, there are certain enzymes which make the cell to repair or eliminate the genetic damage. The damages caused due to ultraviolet rays are repaired by activation of an enzyme upon exposure of dimers—having DNA to strong illumination. These damages could also be repaired in dark in which there occurs the removal of a part of DNA carrying the dimers through the activity of enzymes.

Ionising radiations

The radiation concerning pollution is ionizing radiation, which is the radiation of sufficiently great energy to ionize atoms and molecules. An atom gets ionized when it gains sufficient energy for one or more of its electrons to get separated from the atom; ionization of a molecule might yield two charged fragments, such as $H_2O \rightarrow H^+ + OH^{-*}$. If the fragments are uncharged, then they are referred to as "free radicals," as in the example $H_2O \rightarrow H + OH$. Free radicals are generally very reactive chemically. The most important types of ionizing radiation from the standpoint of pollution (as opposed to the standpoint of the scientific researcher interested in new knowledge of the fundamental properties of matter) have been alpha, beta and gamma radiation:

Alpha radiation is made up of energetic alpha particles. An alpha particle is a He^4 nucleus, which has two protons and two neutrons which are bound together into a very stable particle by the "strong nuclear interaction". Alpha particles carry a positive electric charge whose absolute magnitude has been twice that of the electron's charge and have been capable of interacting strongly with ordinary matter (including living tissues) by electromagnetic interactions.

Beta radiation is made up of energetic electrons (or their antiparticles, the positions). These particles are produced in the beta decay of nuclei, which also produces—antineutrinos (or neutrinos) which interact only very weekly with matter and have been of no importance as radiation As beta particles being electrically charged, they can interact strongly with matter by the electromagnetic interaction.

Gamma radiation are made up of very energetic photons, i.e., very short wavelength electromagnetic radiation. Despite being uncharged, photons are able to bring about very strong electromagnetic interaction with matter. Also, many other types of ionizing radiation are encountered less frequently, but they can also be dangerous to living systems—protons, neutrons, deuterons, and other nuclei of sufficiently great energy, for example. the particular importance of alpha, beta and gamma radiation lies in the fact that they are produced in the radio-active decay of certain atomic nuclei, i.e., in the spontaneous or induced disintegration of these nuclei.

Just as heat and light transfer energy from the sun to the earth and the atmosphere, so does nuclear radiation transfer energy from a source to the absorbing medium. (The source of nuclear radiation may be radioactive atoms or equipment such as X-ray machines.) the effect of absorbing the more familiar types of radiation, such as heat, has been to raise the temperature of the sbsorbing medium.

If this medium has been the human body, or part of it, the rise in temperature has been sensed, and, if it becomes excessive, avoiding action can be taken, by sheltering under a sunshade (shielding), for example, or by moving further away from a fire (distance). A dose of gamma or other nuclear radiation large enough to be lethal to a human being would be able to increase the body temperature be less than one-thousand of one degree centigrade.

The body is, therefore, unable to sense even very high intensities of these types of radiation. Nuclear radiation is different from heat and other types of radiation in that each particle or photon has a sufficiently high energy to cause ionization. The high energy is due to the very high velocity of the particles or the short wavelength of the X and Y- radiation.

Ionization means the removal of an orbital electron from an atom. As the electron has a negative charge, the atom is consequently left positively charged. The atom and the electron, so separated are called an ion pair, that is, a positive ion (the atom) and negative ion (the electron). The absorption of radiation in medium results, then, in the production of ion pairs in the medium. A small amount of energy is needed to bring about ionisation so that in producing ion pairs, the particles or photons of radiation would lose energy of the medium. Normally positive and negative ions recombine to form neutral atoms and the energy originally given to the ion pair gets converted into heat energy.

If the absorbing medium has been a gas, such as air, the ions can be prevented from recombining by applying an electric field. This is carried out by applying a voltage between two plates (electrodes) with a gas gap between them. An example of such a system, is an ionization chamber, in which ion pairs are being produced along the track of an alpha-particle. If the applied voltage is sufficiently high, negative ions produced in the volume between the electrodes get attracted to the positive electrode and positive ions to the negative electrode.

The flow of ions to the respective electrodes constitutes an electrical current and because this is proportional to the intensity of radiation, ion chambers provide a means of measuring radiation. It should be realized that in the case of beta-particles, several hundreds of ion pairs are formed per centimetre of track in air, and in the case of alpha particles some tens of thousands. In media like water (of which the human body is largely composed) ionization can give rise to breakdown of water molecules and the formation of chemical forms which are damaging to biological material. The harmful effects of radiation on the human system, have been largely attributable to such chemical reactions.

As already mentioned, the ionization of a gas offers a means of detecting radiation and the first widely used radiation unit, the roentgen, was based on the ionizing effect on air of X-and Y^- radiation. Intensity of emission may be defined as the amount of radiation passing per unit of time through a unit of surface normal to the direction or radiation.

Intensity is measured in the following units: erg cm^{-2} Sec^{-1}, MeV cm^{-3} sec^{-1}, volt cm^{-2} sec^{-1}. Intensity describes only the incident energy of radiation and says nothing about the quantity of radiant energy absorbed by a medium. In order to evaluate the latter quantity, the concept of a dose of radiation, R is introduced. Dose is the quantity of radiant energy absorbed per unit of volume of irradiated substance. The quantity absorbed by a unit volume per unit of time is called the dose rate of radiation.

A number of units have been devised to describe various properties of X-rays, gamma rays and streams of particles of automatic radiations. Ionizing radiation, except in rare circumstances, cannot be seen, heard, tasted, smelled, or felt. The intensity of radiations, therefore, cannot be measured directly. It is measured, instead, by the degree of molecular or atomic disruption (ionization) when the radiation passes through a specified medium. The amount of radiation may be expressed in terms of (a) ionization produced in a given quantity of air (roentgens), (b) energy absorbed by tissue (rads), or (c) degree of biological effect estimated to occur in the body (rems).

It may be defined as the number of radioactive disintegrations occurring each second for a given amount of radioactive material. This equals Nl, where N is the number of atoms present and l is the radioactive decay constant giving the probability of decay of the atom in question. For $Ra^{2:6,}$ the naturally occurring isotope of radium, l = 1.36 x 10^{-11}/s, so 1 g of Ra^{226} (which has 6.025 x 10^{23}/226 = 2.7 x 10^{21} atoms) will produce 2.7 x 10^{21} x 1.36 x 10^{10} disintegrations per second.

It may be defined as the quantity of any radioactive material that gives 3.70 x 10^{10} disintegrations per sound. This unit was defined so that 1 g of natural radium together with its decay products amounted to 1 curie. Millicurie and microcurie units are also used. The in frequently encountered Rutherford is defined to be exactly 1 million disintegrations per second. The curie is a unit of activity, not of quantity of radioactive material

1 Millicurie (mCi) = 10^{-3} Ci

1 Microcurie (Ci = 10^{-6} Ci

1 Picocurie (pCi) = 10^{-12} Ci

The SI unit of activity is the becquerel (Bq).

1 Bq= 1 disintegration or other nuclear transformation per second

1 Ci = 3.7 x 10^{10} Bq.

Originally roentgen was defined as the amount of gamma (or X) radiation that was produced by ionization of 1 e.s.U. of electricity of each sig~ in 1 cm^3 of dry air at 0^0C and atmospheric pressure; equivalently it produced

2.08 x 10^9 singly charged ions of each sign. At one time the roentgen was believed to correspond to the absorption of 83 ergs/g of air since it was thought that 32.5 eV were required, on the average, to produce one ion-pair; later, 86 ergs/g of air was determined to be a more accurate value. Experiments showed that 1 roentgen of gamma radiation led to the absorption of about 97 erg/g in soft tissue, and a new unit was defined : The rep (for "roentgen equivalent physical") denoted a dose of 97 erg/g of body tissue.

To give an idea of magnitude, if each person in a large population were exposed to 650 roentgens, half would die in 1-4 weeks. The Roentgen (R) is a unit of radiation that is useful only for x-rays and gamma rays and is the standard measure of radiation used by radiologist. The designation is named after the German physicist Wilhelm Roentgen who discovered x-rays.

According to the definition adopted by the International Radiological Congress of 1937 the roentgen is the quantity of x or gamma-radiation at which the corpuscular emission (electrons) associated with it, produces, in air, ions carrying one electrostatic unit of quantity of electricity of either sign per 0.00129 g of dry air (where 0.00129 g is the mass associated with 1 cm^3 of dry air at 760 mm Hg).

Although it has been still in use, the roentgen is inadequate as a radiation unit because it applies only to X and λ-radiation and their effect on air. Usually, human tissue is the medium, of interest and the energy deposition is often higher in tissue than in air. In the case of λ-radiation, over the commonly encountered range of photon energy, the energy deposition in tissue for an exposure of 1 R has been abut 0.0096 JK/g. The concept of radiation absorbed dose was introduced to overcome these difficulties.

Since the roentgen was designed to be used only with x-rays and gamma- rays, it was necessary to device a unit that could be used to specifically define the dosage from alpha and beta particles and fast neutrons as well as other kinds of radiation. The unit selected, called the rad (radiation absorbed dose), is now the most widely designation for radiation dosage. For practical purposes, the rad is nearly the equivalent of the roentgen. Rad may be defined as that quantity of radiation that leads to the absorption of 100 ergs/g of the absorbing material.

A given quantity of radiation will correspond to different numbers of rads in different materials but the number of roentgens will be the same in each case. In radiation biology the dose is the energy released in tissue by ionizing radiations.

1 rad = 10^{-5} joule energy absorbed per gram of tissue

= 0.01 joule per kilogram

1 millirad (mrad) = 10^{-3} rad.

A total body-absorbed dose or about 6000 rads of gamma radiation would be lethal for most people. The roentgen and the rad are quite similar in size. One roentgen of X or gamma-radiation will deliver very nearly one rad when it is absorbed by muscle tissue.

The SI unit of absorbed dose is the gray (Gy).

1Gy = 1 joule energy absorbed per kilogram of tissue

Absorbed dose is a measure of energy deposition in any medium by all types of ionizing radiation. The unit of absorbed dose is the rad which may also be defined as an energy deposition of 0.01 J/kg.

Since an exposure of 1 R results in an energy deposition of 0.00869 J/kg in air and 0.0096 J/kg in tissue, it follows that 1 R gives an absorbed dose of:

$$\frac{0.00869}{0.01} = 0.869 \text{ rad in air}$$

and

$$\frac{0.0096}{0.01} = 0.96 \text{ rad in tissue.}$$

In many cases, then, exposure in R and absorbed dose in rads have almost the same value. Note that the medium has been always specified.

Although the rad is a very useful physical unit, it transpires in biological system the same degree of damage is not necessarily produced by the same absorbed dose of different types of radiation. It is found, for example, that 1 rad of alpha radiation can do as much biological damage as 10 rad of gamma radiation.

This differences in the radiobiological effectiveness must be taken into account if we wish to add doses of different radiations to obtain the total biologically effective dose. To do this we must multiply the absorbed dose of each type of radiation by a quality factor (QF) which reflects the ability of the particular type of radiation to cause damage.

The quantity obtained when absorbed dose is multiplied by quality factor is known as the dose equivalent, the unit of which is the rem.

Dose equivalent (rem) = Absorbed dose (rad) x QF

The value of the quality factor has been found to depend

on the density of ionization caused by the radiation. And alpha particle produces about 10 million ion pairs per centimeter of track in tissue whereas a beta particle produces about one hundred thousand per centimeter.

Table 1

Summary of Values of QF

Type of radiation	*QF*
X, γ and β	1
Thermal neutrons	3
α and fast neutrons	10

Gamma radiation causes ionization of a similar density as beta radiation and since all other types are compared to gamma, the QF for gamma rays is one and the QF for beta is also one. For alpha particles. The QF is taken as 10. The QF for neutrons is dependent on the neutron energy, but a value of 3 is usually taken for thermal neutrons and 10 for fast neutron. The values of QF are summarized in Table 1.

Example: In one years a worker receives a gamma dose of 2 rad, a thermal neutron (N_s) dose of 0.5 rad and a fast neudose (N_f) of 0.1 rad. What is his total dose equivalent?

Dose equivalent = Absorbed doss X QF

dose equivalent = 2 x 1 = 2 rem

N_s dose equivalent = 0.5 x 3 = 1.5 rem

N_f dose equivalent = 0.1 x 10 = 1 rem

Total dose equivalent = 4.5 rem

When calculating the dose equivalent due to radioactive material inside the human body, it is necessary in certain rare cases to apply a further factor which takes into account nonuniform distribution of the radioactivity. In such cases the dose equivalent is expressed:

Dose equivalent = Absorbed dose x QF x DF

where DF is a distribution factor. Normally, the value of DF is one, but in the rare cases mentioned it is given a value of 5. Rem may also be defined as that quantity of radiation that produces the same biological damage in man as one rep of gamma radiation. Different types of radiation seem to have different biological effects, and they can be characterized by their "relative biological effectiveness" (RBE), which gives the number of rems per rad. The RBE really depends on the effect considered, the location of the radiation source, and other things, but typical values are 1 for gamma and beta radiation of all energies above 50 keV, 2to 5 for lower energy gamma and beta radiation of all energies above 50 keV, 2 to 5 for lower energy gamma and beta radiation, 10 for fast neutron and protons (0.5 to 10 MeV) and natural alpha particles, and 20 for heavy nuclei and fission particles.

Doses expressed in rads are not necessarily additive; the nature of the radiation and other factors influence the effect on the irradiated body. If an organ receives 5 millirems (mrem) of neutron radiation and 10 mrem of gamma rays, it has taken 15 mrem of total dose equivalent. Roughly, however, rads, and roentgens are nearly equivalent in comparing dangers to human health and life.

The rem was intended to compensate for the properties of the wide varieties of radiation unaccounted by relating them to the effect of electromagnetic radiation. The rem is an estimate of the amount of radiation of any type which produces the same biological injury in man as that resulting from the absorption of a given amount of X-radiation or gamma radiation.

The dose unit, roentgen cannot be used for measurement of corpuscular emission. There are also difficulties. As we have seen above, in its application for the determination of absorption in various substances. This led to the introduction of another measure of the quantity of absorbed energy—the roentgen equivalent physical or rep. Numerically, the rep is equal to the energy equivalent of the roentgen, in irradiation of water (93 ergs/g). In contrast to the Roentgen, rep is a constant quantity of energy absorbed per 1 gram of substance, regardless of the type of radiation and the chemical constitution of the substance.

$$1 \text{ rad} = 1.19 \text{ rep} = 6.25 \times 10^{13} \text{ eV/g}.$$

For the sake of brevity, the rep will, in most cases, be indicated as R (roentgen). The most recent (1980) Systemel International (SI), unit is gray (abbreviated Gy) which is one joule per kilogram. Since 1 joule = 10^7 ergs, and 1 rad = 100 ergs/g, 1 gray = 100 rads.

The greatest difficulties are encountered in the determination of the absorbed energy and in the choice of units for its measurement when the radiation from atomic reactor is involved. In this case, one deals not with a homogenous stream, but with a mixture of slow (thermal) neutrons, their accompanying gamma-rays, as well as fast neutrons all of these in varying ratios. The problem is complicated not only because the thermal neutron absorption cross-section differs for the individual elements of the polymer, but also because the ratio of the comments of the emission varies from reactor to reactor.

The unit nvt has been often used for measurement of a neutron stream. This unit is the product of neutron density (thermal neutrons/cm^3, neutron velocity (cm/sec) and irradiation time (sec). Its dimensions are the number of neutrons/cm^2.

Electron Volt (eV) is an extremely small unit of

energy use to describe the energy of particles in a stream of radiation.

1 electron volt $= 3.8 \times 10^{-20}$ calorie

$= 15.9 \times 10^{-20}$ joule $= 0.159$ attojoule (aJ)

(It comes from the energy an electron gains when it accelerates under a potential difference of 1 volt in a vacuum).

1 kilo - electron- volt (keV) = 1000 electron-volts

1 meg-electron-volt (MeV) = 1,000,000 electron-volts

The gamma rays from a cobalt-60 source used in cancer therapy have energy on the order of 1 MeV. Alpha particles from radium have about 5 MeV energy. X-rays used to diagnose disease have energies on the order of 0.09 MeV (90 keV). Primary cosmic rays sweep into our upper atmosphere with energies ranging roughly from 200 MeV to 200 "BeV" (billion electron-volts, or properly, giga-electron-volts, Gev).

Linear Energy Transfer (LET)

It may be defined as the energy given up by radiation per unit path length as it goes through air, tissue, or other matter. Alpha particles and neutrons are high LET radiations. Gamma and X-rays generate lower LET paths in tissev.

Half-Life

Any radioactive substance decays away in such a fashion that dual time intervals which reduce it by equal proportions. The time interval which reduces the quantity by half is the 'half life' and is a characteristic of the particular substance.

In symbols,

$$Nb/Na = 2(tA\text{-}tB)t1/2$$

where Na and Nb are the number of nuclei of a substance present in a sample at times tA and tB respectively, and t is the substance's half-life. From this formula one can see that when the time interval tB - tA is equal to the half-life then the ratio NP/NA is equal to.

When the interval is twice the half-life, the ratio is a quarter; when the interval is three half lives, the ratio is an eighth; when the interval is four half-lives, the ratio is a sixteenth; and so on. Instead of the half-life, one sometimes uses the 'mean life' or average life-denoting it by tM we have

$$NB/NA=\exp\ [tB\text{-}tA)tM)]$$

and

$$t^1/_2 = tM.\ In2\ (In2 = 0.693)$$

For example, the half-life of the neutron is 11 minutes, and so the average life-time of a neutron is about 15 minutes: of any sample of neutrons, 99.9 will have decayed after an interval of ten half-lives, i.e. 2 hours. See also exponential.

LD_{50}/30-Days

It may be defined as the 30-day medium lethal dose equivalent, the dose equivalent (in rems or millirems) that will kill 50% of the exposed individuals within 30 days. It is estimated to be 500 rems for man and it is known to be 300 rems for dogs, 600 rems for mice and 700 rems for rats.

Sub-multiples

In terms of occupational radiation exposure the roentgen rad and rem are rather large units. It is often convenient to have smaller units and this is done by using the prefixes milli (one thousandth) abbreviated to m, and micro (one millionth) abbreviated to μ. Thus :

1 roentgen (R) = 1000 milliroentgen (mR)
= 1000000 microroentgen (μR)

1 rad = 1000 millirad (mrad)
= 1000000 microrad (μrad)

1 rem = 1000 millirem (mrem)
=1000000 microrem (μrem)

Example: On 3 successive days a nuclear reactor operator receives the following gamma dose-equivalent:

Day 1, 9.5 mrem; Day 2, 500 μ rem; Day 3, 0.1 rem

What was his total dose equivalent in mrem over the 3 days ?

Day	*Dose equivalent*	*mrem*
1	9.5 mrem	9.5
2	500 μsem = ——— μmrem	0.5
3	0.1 rem = 0.1 x 1000 mrem	100
	Total doze equivalent	110 mrem

Dose Rate

The roentgen, rad and rem have been units expressing an amount of radiation which may have been received over any period of time. In controlling the radiation hazard it is usually necessary to know the rate at which the radiation is being received.

If a man works in an area for 2 hours and receives a dose equivalent of 0.4 rem, then the dose equivalent rate is 0.2 rem per hour (rem/h). Similarly, we can have exposure rates (R/h) and absorbed dose rates rad/h). The relationship between dose, dose rate and time is:

Dose = Dose rate x time

The use of the three units, the roentgen, the rad and the rem is unfortunate and, to the newcomer to health physics, confusing. Whereas exposure, in roentgens, can be used only to describe the effect of X and γ-rays on air, the rad and rem are applicable to all types of radiation.

The rad describes an absorbed does in any medium and the rem expresses biological effect on the human body. In health science, it is clearly the biological effect of radiation which is of interest and so whenever possible dose equivalent in rem should be the unit used. In everyday health science, the term dose is often loosely used to mean any one of the quantities exposure, absorbed dose or dose equivalent.

Radiation Sources in the Environment

Throughout history man is exposed to radiation from the environment in which he lives. This natural background radiation comes from three main sources-cosmic radiation, radiation from terrestrial sources and radioactivity in the body. It has been impossible to decide whether the natural background radiation has been harmful or beneficial to the development of the human species. It is known that a very small, but finite, fraction of the natural mutations in cells must be beneficial as it contributes to the evolution of higher forms of life.

On the other hand, a major proportion of all genetic mutations gives rise to hereditary defects and genetic of death. It is clear that these two effects have achieved some sort of balance and that man has evolved to his present state in spite of background radiation, or perhaps even because of it.

In addition to the natural sources of background radiation many artificial sources of radiation have been introduced in the past 70 years. These artificial sources now add a significant contribution to the total radiation exposure of the population.

Naturally and Artificially Occurring Radioactive Elements

Radiation sources in the environment has been partly natural and partly man-made. Some natural radiation is produced by naturally occurring radioactive elements, such as the ones listed in Table 2.

The table includes the list of the radioactive isotopes, their approximate abundances in the lithosphere (the earth's crust), their half-lives and the nature of their radiation. These constitute an important source of radiation exposure to humans. The amount of whole body radiation produced annually has been found to vary from place to place, depending on such variables as the nature of the soil, and the distance to material deposits. Many places are known in Brazil where the soils are so rich in thorium and uranium that whole-body exposures can amount to 2 rem annually, perhaps 40 times the average U.V. radiation exposure from earth.

There are some natural radioactive isotopes (K^{40}, C^{14}, Th^{232}, U^{238}, etc.) which are present in human body tissues themselves, the amounts varying somewhat with geographical location, type of water supply, and other variables. Some naturally occurring radioactive isotopes, their abundances in the lithosphere, their half-lives, and the nature of their radiation, Ra^{216}, U^{238} and Th^{232} are all members of natural radioactive series that produce all three types of radiation-alpha, beta, and gamma.

Table 2

Isotopes	*Abundance*	*Half-life (years)*	*Radiation*
Ra^{226}	$2x10^{-12}$ (in soil)	1622	Alpha, gamma
U^{238}	$4x10^{-6}$ (in soil)	$4.5x10^{9}$	Alpha

Th^{232}	12×10^{-6} (in soil)	1.4×10^{10}	Alpha, gamma
K^{40}	3 ppm	1.3×10^{9}	Beta, gamma
V^{50}	0.2 ppm	5×10^{14}	gamma
Rb^{87}	75 ppm	4.7×10^{10}	Beta
In^{115}	0.1 ppm	6×10^{14}	Beta
La^{138}	0.01 ppm	1.1×10^{11}	Beta, gamma
Sm^{147}	1 ppm	1.2×10^{11}	Alpha
Lu^{176}	0.01 ppm	2.1×10^{10}	Beta, gamma

The rocks and soil of the earth's strata contain small quantities of the radioactive elements uranium and thorium with their daughter products. The concentration of these elements varies considerably depending on the type of rock formation. In sandstone and limestone regions the concentration is much lower than in granite.

The dose rate from this source depends on the geographical location. In the British Isles the dose rate to the whole body varies between about 30 and 60m rem per year. Other long-lived isotopes such as ^{48}Ca (>77 x 10^{8} years) ^{50}V (4 x 10^{4} years), do not contribute significantly to human dose because they occur in nature in very small concentrations.

Plutonium

Plutonium, chemical symbol Pu, is an artificially produced, radioactive, chemically and radiologically highly toxic metallic element in the actinide series, which also occurs in minute quantities as a result of natural disintegration processes in nature. It is formed in nuclear reactors from uranium (U) by the following process: U-238 +neutron → U-239; U-239 disintegrates by the emission of beta radiation into neptunium-239 (Np-239), and this decomposes in the same way into Pu-239.

Plutonium has the atomic number 94; and from it the isotopes Pu-239 to Pu-246 are known, of which Pu-

244, with 8.2 x 10^7 years, has the longest half-life. The other isotopes also have a very long-life. The technically most important of them, Pu-239, has a half-life of 24,400 years.

As it is highly fissionable by slow neutrons, Pu-239 is used as nuclear fuel and to make atomic bombs (the atomic bomb that was dropped on the Japanese city of Nagasaki on August 9, 1945, was a plutonium bomb). It is easier to extract than uranium (U-235), and its critical mass amounts to no more than 8-16kg, compared with a critical mass for uranium of 50 kg.

Pu-239 is now obtained in large quantities from nuclear reactors, especially from fast breeder reactors. A fast breeder contains 1 ton of plutonium its fuel rods. Radiologically, plutonium, which burns in the air to form aerosols capable of penetrating into the lungs, is one of the strongest poisons known.

When they get into the human organism, quantities of as little as one ten-millionth part of a gram are sufficient to cause cancer, for plutonium emits alpha rays of limited range which are capable of very high radiation doses locally. Investigations by Tamplin and Cochran in 1974 serve to demonstrate that the margins for plutonium hitherto considered tolerable, and the legal limits prescribed by law for plutonium contamination, are in fact too high by a factor of 300,000. Plutonium, is given off into the environment in limited quantities by reprocessing of nuclear fuels.

There is likely to be in the future a substantial increase in the transport of plutonium to reprocessing centres by road and rail, so that the possibility of releases of plutonium by leakage or owing to terrorist activities cannot be ruled out. If a transport vehicle containing 25 kg were to overturn in an accident, the plutonium so released could cause up to 250 billion cases of lung cancer.

From the same quantity of plutonium, given sufficient technical knowledge, an atomic bomb could be manufactured. The dangerous nature of plutonium, having regard in particular for the long life-span of its isotopes, is one the main arguments against using nuclear fission as a source of energy.

Radioactivity in the body

The human body contains very small quantities of the radioactive isotopes carbon-14 and potassium-40. The ^{14}C originatesin the atmosphere and results in a dose of one mean per year in the soft tissue. Potassium 40 (40K) is naturally-occurring (half-life 1.27×10^9 years) and contributes about 20m rem per year to the gonads. A significant contribution to the radioactivity in the body comes from the gaseous decay products of the uranium and thorium radioactive series, namely radium and boron. these gases diffuse from the rocks and soil and are present in easily measurable concentrations in the atmosphere. They are breathed by man along with their decay products and are also taken up by plants and animals with the result that most foodstuffs contain measurable amounts of natural radioactivity.

Of ordinary foods, cereals have a high radioactive content while milk produce, fruit and vegetables have a low content. The intake of this type of radioactivity varies greatly with diet and is thought to contribute an average gonadal dose of about 4 mrem per year in the British Isles.

Cosmic Radiation

Another important source of natural radiation has been cosmic radiation. Cosmic rays have been high-energy charged particles (mostly protons) of extra terrestrial origin. They have been capable of producing other energetic radiation by collisions with oxygen and nitrogen nuclei in the atmosphere. Table 3 lists some products of cosmic rays.

The radiation dose obtained from cosmic rays has been found to be greater at greater heights in the atmosphere and may amount to about 10 rem/year above the atmosphere. Natural radiation from cosmic rays will be higher at higher elevations and higher in a jet aircraft at 10 km than on the ground.

Products of cosmic radiation, their half-lives, and their concentration in disintegrations per minute per cubic meter of air in the lower troposphere. The concentrations refer to those resulting from cosmic radiation and do not include the effects of nuclear weapons tests.

Table 3

Isotope	*Half-life*	*Concentration*
H^3	12.3 years	10
C^{14}	5760 years	4
Be^7	53 days	1
S^{35}	87 days	0.015
P^{32}	25 days	0.015
P^{31}	14.3 days	0.02

Cosmic radiation reaches the earth from inter stellar space and from the sun. It is composed of a very wide range of penetrating radiations which undergo many types of reactions with the elements they encounter in the atmosphere.

The atmosphere acts as a shield and reduces very considerably the amount of cosmic radiation reaching the Earth's surface. This filtering action means that the dose rate at sea-level is less than at high altitudes. For example, the mean dose rate to the gonads at sea-level, at the equator, has been measured as 23 mrem/year, while the equivalent dose rate at 10,000 feet, was 56 mrem/year. The average dose rate in the British Isles from cosmic radiation is about 50 mrem/year.

One very important radionuclide arises from the interaction of neutrons in cosmic radiation with nitrogen in the upper atmosphere to carbon-14 as follows:

$$^{14}N(n, p)^{14}C$$

The carbon-14, which has a half-life of 5568 years' diffuses to the lower atmosphere where it may become incorporated in living matter. Similarly, small concentrations of other radionuclide such as ^{3}H (half life 12.26 years) ^{36}Cl (half life 3.08 x 10^5 years) and ^{41}Ca (half-life 1.1 x 10^5 years) are maintained in the lower atmosphere by cosmic ray reactions. They are much less important than ^{14}C.

Background Radiation

Natural atomic radiation emitted by the trace concentrations of radioactive materials in the rocks and soils of our planet as well as the cosmic rays coming from the sun and outer space. Levels of background radiation are usually given in millirems (mrem) per year, the unit of dose equivalent, but sometimes the nearly equivalent unit of the millirad (mrad) per year is used.

Table 4

Background Radiation

Background radiation	*AmountVoluntary (mrem/year)*	*Amount exposure*	*(mrem/year)*
Cosmic rays	50	Jet flight, cross	
Ground (6 hours/day)	15	country	4
Building materials		Watch dials	2
(18 hours/day)	45	Television	
		(1 hour/day)	5
Air	5	Medical and	
		diagnostic	
Food and water	25	X-ray	55
Total	140	Total (est)	65

Source:A.P.Bray.

Although the background radiation varies widely from place to place in the world. For most settled areas of the United States it averages 140 mrem/year. In the Travancore region of eastern India, active monazite sands produce a background of 200-2600 mrem/year. Certain states in Brazil have the same sand, the background radiation there ranges from 500 to 1500 mrem/year.

About 50 of the approximately 350 isotopes of naturally occurring elements are radioactive. On an average, the top 6 inches of soil on our planet contains a gram of radium per square mile. One of the "daughters" of radium, created as radium decays, is the radioactive gas, radon. Granitic rocks contain radioactive materials. (In fact, granite may someday be processed to obtain nuclear fuels).

We take in radioactive materials in the air we breathe, the food we eat, and the fluids we drink. Total Exposure of the Average Person in the United States to Ionizing Radiations. An individual's exposure to radiations in excess of background comes from watching television, travelling in a high altitude jet (giving more exposure to cosmic rays,) wearing a watch with a luminous dial, and being X-rayed at the office of the dentist or the doctor.

These more recent additions to our total exposure to radiations are part of the reason for the concern of many radiation biologists. Moreover, nuclear power plants add radioactive materials to the water, the air, and the food chain. A nuclear power plant run without incident may add 5 mrem/year to the atmosphere and 0.05 mrem/year to the water used for cooling within the plant. In relation to the background radiation, this has been judged to be insignificant unless one ignores the concentrating effect of food chain on certain radioactive isotopes or the cumulative effect of several nuclear plants clustered in densely populated areas.

Man-made Sources of Radiation

The early experiences of man-made sources of radiation involved X-rays and various uses of radium. As early as 1896 a letter appeared in Nature describing the effects of repeated exposure of the hands to X-rays and during the next 15 years many more cases were reported. These cases arose both from experiments with X-ray sets and also from their use in various treatments.

By 1911, Hesse had studied the histories of 94 cases of tomorrows induced in man by X-rays, of which 50 cases were among radiologists. These studies illustrated the early types of damage produced by X-rays but gave no indication of the longer-term effects.

For some types of damage, such as skin cancer, there is a latent period of between 10^{-30} years and some radiologists observed malignant skin changes as late as 25years after discontinuing fluoroscopic examinations. By 1922 it was estimated that more than 100 radiologists had died from occupationally-produced cancer. Similarly, it has been estimated that the death-rate from leukaemia among early radiologists was about nine times that among other physicians.

Other studies have shown that the average life-expectancy of the pioneer radiologists was reduced by approximately 203 years compared with physicians in general practice. Using results from animal experiments on the relationship between lifeshortening and amount of irradiation, it is estimated that the total dose received by the average radiologist from 1935-1958 ranged between 400-800R.

The results of this study have been used by the International Commission on Radiological Protection (I.C.R.P.) to set the occupational exposure limit of 15 rems per year for most organs of the body. Experience was also gained early in this century of the effect of internal dose from various nuclides such as radium (^{226}RA, half-life 1622 years), mesothorium (^{228}Ra, half-life 5.8 years), radiothorium (^{229}Th, half-life 1.91 years), and their daughter products.

Even earlier, it had been recognized that there was a remarkably high incidence of lung cancer among the minors in the Schneeberg cobalt mines of Saxony and the Joachimsthal pitchblende mines in Bohemia. Eventually, this high rate of lung cancer was shown to be due to radiation from the daughter products of uranium, namely ^{226}Ra, ^{222}Rn, ^{218}Po etc. These mines contain large concentrations of uranium.

During the second and third decades of the twentieth century there were many cases of overexposure to radium. A considerable number of these arose from the use of radium as a therapeutic agent. It was administered for a large variety of diseases ranging from arthritis to insanity! The most serious overexposures to radium occurred in the radium-dial painting industry in the United States. Most of the persons employed were women and they had the habit of 'pointing' their paint brushes with their lips.

Many of these women probably ingested tens or even hundreds of microcuries of radium. It is not known how many radiumdial painters actually died from the effects of radiation damage. The maximum permissible body burden of ^{226}Ra set by the International Commission on Radiological Protection (I.C.R.P.) is 0.1 Ci which corresponds to an average dose to the skeleton of about 30 rem/year. The study of people exposed to X-rays and radium is continuing to try to establish the degree of risk from acute or chronic exposures to radiation.

Radiations from Medical and Dental Exposure

The public are getting a slowly increasing dose of

ionizing radiations from medical and central X-rays. As Table 5 indicates these man-made exposures are considered the largest source, above the background, of ionizing radiations of man. Most of the exposure has been for diagnosis ("checkups").

Unfortunately, most of the X-ray machines used for medical purposes have been operated by physicians and chiropractors who lack special training in health physics, radiation protection, or the use of their machines. Because radiologists, the specialists in medical X-rays work, tend to be employed at hospitals and clinics with high workloads, the majority of X-ray actually obtained are taken under the supervision of radiologists.

The nonspecialist, has been more likely to have old equipment, to understand its use less well, to allow greater exposures than necessary (sometimes because of a cheap camera and a slow lens), and to be less careful in focusing the beam.

These generalizations are also applicable to the field of dentistry as well. There should be laws relating to periodic inspections of equipment and to requtirements for the training of physicians and dentists have been exceedingly lax. One is greatly concerned about pelvic X-rays given to women of child-bearing age. The embryo and fetus have been particularly sensitive to damaging radiations. the International Commission of Radiation Protection and Measurement (ICRP) urged in 1962 that except in emergencies, any radiological examinations of women in this age group be limited to periods when pregnancy is most improbable, which the ICRP defined as the 10-day period following the onset of menstruation.

It has been estimated that 75-90 per cent of the total exposure of the population from medical uses of radiation comes from the diagnostic use of X-rays. The most critical regions of the body are the bone marrow, the gonads and the foetus.

The bone marrow is the site of the primitive bloodforming cells and so irradiation of this region could lead to the induction of leukaemia. Irradiation of the gonads is important because of the possibility genetic damage. Irradiation of pregnant women should be very severely limited because it can lead to physical and mental deformities of the child.

In most countries the average dose to the population from therapeutic radiology is much less than that from diagnostic radiology. Although quite large exposures may be used in certain treatments, only a small number of people are involved.

Uses of Radioisotopes

Radioisotopes are used in medicine to give a means of tracing the path and location of specific chemicals in the body. Since radioactive isotopes are chemically identical to stable isotopes of the same element, they will follow the same path and be concentrated to the same degree as the non active isotopes in the body. By counting methods the location of the active, and hence of the ordinary non-active, isotopes of the element may be determined.

Radiation from Television sets

For a person viewing television at a distance of 6-7 feet, the average absorbed dose to the gonads, according to data quoted by Morgan, is probably much less than 1 millirad/year, provided there is a safety glass in front of the screen and the set operates at less than 25 -kilovotts. For a 25-kilovolt set without the glass, the absorbed doses at 6-7 feet could be as high as 1.1 millirads/year to ovaries and 7.5 millrads/year to testes.

Since for radiation protection purpose the units of the rad and the rem are taken as roughly equivalent, this dosage, as seen in Table 3 exceeds exposures caused to the general population by currently operating nuclear power plants and is of the same order of magnitude as exposures being currently received as the result of fallout from nuclear weapons testing.

This dosage or any widely distributed dosage is a serious matter because of the large mumber of people involved, therefore, the potential changes in the gene pool, as well as the statistical increase in incidence of leukeamia and other diseases induced by radiations.

Radiation from Nuclear Power Plants

The contribution of nuclear power plants to man-made radiation is appreciably small. Radiation standards set by the U.S. Atomic Energy Commission (AEC) formerly allowed a maximum dose of 500 mrem/year at any point on or beyond the boundary of a nuclear power plant, which would several times the usual natural background; in 1971 this limit, got reduced to 5 mrem/year for light-water-cooled nuclear power reactors.

Operating nuclear power plants (except for less than half a dozen older plants) in practice emitted no more than a few per cent of the old 500 mrem/year limit , and in the great majority of cases less than 1% of it.

Hence the actual exposure of a person remaining around the clock at the least favorable location (in the direction of the prevailing winds at the plant boundary) would be no more than 5mrem/year, in accordance with the new limits. It is thus believed unlikely that anyone in the U.S. is receiving even 1 mrem/year radiation exposure from nuclear power plants—an amount which is approximately equal to that received from increased cosmic radiation exposure during a single transcontinental jet flight.

Radioactive Fallout from Nuclear Weapons

Radioactive fallout from nuclear weapons tests has been of grave concern to scientists. When we test a nuclear weapon in the atmosphere, there occurs local fallout of radioactive fission products over the immediate area for about a day, then worldwide tropospheric fallout for about a month from fission products released into the troposphere, and stratospheric fallout worldwise for many years thereafter. This fallout has been easily detected at scientific laboratories which are located around the world and its study has greatly increased our knowledge of worldwide atmospheric transport processes.

The rapid transport of fallout from one hemisphere to the other has been indicated by the detection within 22 days at 34° north latitude of radioactive I131 and Ba 140 from a French nuclear test at 21° south latitude. So as to be of concern to man, a radioactive fission product must be produced in sufficient quantities, possess a sufficiently long half-life (which is as short as 8 days in the case of I^{131}), be transferable to man, and remain in the body long enough to do damage.

The most hazardous radionuclides turn out to be those listed in Table 6, C^{14} has been produced naturally in the atmosphere by cosmic radiation and has been a constituent of all living tissues; weapons tests through 1975 increased the amount in the atmosphere by 80 to 100% but circulation in the biosphere will be able to reduce this to about 3% by 2040A.D.unless atmospheric testing continues. Sr^{89} has been similar to (but much shorter-lived than) Sr^{90} as a pollutant that gets discriminated against biologically but has been still of great importance. Sr^{90} and I^{131} both reach man through cow's milk, the Sr^{90} going into bones and I^{131} into the thyroid gland.

Cs^{137} reaches human tissues through ingestion of milk and meat but has a biological half-life (i.e., half-life inside the body) of only 70 to 140 days because of its fairly rapid removal from the body through metabolic action.

Table 6

Radionuclides Important in Fallout

Element	*Isotope*	*Half-Life*
Carbon	C^{14}	5760
Strontium	Sr^{89}	51 days
Strontium	Sr^{90}	28.9 years
Iodine	I^{131}	8.1 days
Cesium	Cs^{137}	30.2 years

Due to the variations in half-lives (both physical and biological), and environmental happenings that flush materials into unpopulated domains (e.g., the ocean), these averages have been not very meaningful except that they represent doses, in comparison with those received from medical and dental exposures, that are small. (The current average per year is estimated as about 1.5 millirads).

Further atmospheric testing, of course, would have pushed these figures higher and higher. And people near to or downwind from the tests—to say nothing of the victims of Hiroshima and Nagasaki—had been seriously harmed.

Peaceful or Constructive Uses of Nuclear Explosions

Another expected man-made radiation source in the future would be the "peaceful" or "constructive" uses of nuclear explosives for releasing natural gas from underground regions, for building harbors, or even for constructing a new sea level canal between the Oceans.

Natural gas released with a nuclear explosion has been likely to get contaminated by radioactive gases such as Kr^{85}, for example, and large-scale construction processes may lead to radio-active contamination of air and water. Even the use of radio-isotopes for medical and scientific purposes involves some risk to the general population; the burnup of an orbiting SNAP (Systems for Nuclear Auxiliary Power) generator in the atmosphere in April 1964 led to detectable amounts of Pu^{238} in rainwater. The risks involved in the peaceful uses of nuclear energy are outweighed by the benefits. Danger of radioactive pollution principally lies in nuclear war and in testing of nuclear weapons. Both short and long term effects have far reaching implications.

Since 1945, with the initiation of nuclear weapon testing and the disastrous results at Hiroshima and Nagasaki there has been considerable awareness of nuclear hazards. The US nuclear power submarine, lost in the Atlantic in 1963, releases a large quantity of radioactive materials in the environment.

As such, even if nuclear weapons are not used during major wars, radioactive pollution of the ocean is a continuous threat to mankind. the total number of nuclear detonations up to the end of 1978 was 1,619 by the United States of America, 247 by the USSR, 29 by France, 24 by Peoples Republic of China and one by India (c.f Office of Public Affairs, USA). In the United States, it is true that even at the time of maximum testing the average fallout was 30 mrem, well below the guidelines of 170 mrem per year.

But in spite of this minimal dosage the shortening of life span due to any exposure to radiation cannot be over looked. The nuclear tests may be underground, underwater and atmospheric. The underground tests, as carried out by India, can be fully contained and generally result in cratering. The Plowshare Programme or Atomic Energy Commission,USA, recorded that cratering explosions may release 10% of the radioactivity into the atmosphere the ecological balance is usually not upset the danger is posed by the remaining 90% which may be leached and affect the underground water.

It is quite likely that underground explosions may release radioactive contaminants as often noted in the burning of radioactive natural gas in the process of stimulation of its production from the geological strata of low permeability. The classification of the underwater test is based on the criterion of depth under the surface of water and that of the atmospheric test on different heights from the surface of the earth up to the stratosphere. The principal threat to our ecosystem and human life lies in underwater and atmosphere tests. Several peaceful uses of nuclear explosion have been programmed, including the acceleration of gas and oil production and formation of under ground storage chamber for gas and oil.

Others involve use in mining projects, refineries, excavations as well as in the construction of harbours and canals. Precautions are normally taken in such cases to contain hazards up to the level of zero discharge. Though paradoxical, the techniques of radiation and problems of industrial pollution.

Nuclear Wastes

The increasing use of radioisotopes and, more particularly, the development of the nuclear power industry results in an evergrowing quantity of radioactive waste. Continued dispersal of low and intermediate levels of radioactive waste to the environment means that members of the general population will receive an increasing exposure from this source. At present the contribution to the total exposure of the population from waste disposal is negligible but it seems certain to increase as more power stations are built. For this reason, very strict control is exercised over the dispersal of radioactive waste to the environment. Nuclear wastes from atomic power plants comes in the form of spent fuel roads of uranium and the by-products such as plutonium. It is estimated that these can remain toxic to humans for over 200,000 years.

Radioactive iodine, another waste product from power plants can cause cancer of the thyroid gland. Waste coming from the production of nuclear weapons produces radioactive strontium and cesium both of which are carcenogenic. These materials generate heat and penetratin radiations from centuries. The nuclear wastes such as contaminated dust, debris, clothing, industrial clothing trash, etc.

When dumped into the disposal sites can defile the atmosphere and even seep into the soil and pollute the drinking water. It is, therefore, necessary, that wastes coming from power stations and defence establishments should be carefully handled, isolated, buried and protected.

Fallout from Weapons Testing

The nuclides of concern in radioactive fallout from nuclear weapons testing are similar to those arising from the operation of nuclear power stations. The most important radionuclides are strontium-90 (^{90}Sr, half-life. 28.8 years) and cesium-137 (^{137}Cs, halflife 30.0 year). Strontium-90 concentrates in the skeleton and cesium-137 is distributed uniformly throughout the body.

Some of the radionuclides created during a nuclear test are injected into the troposhere (40,000-60,000 feet) and are carried around the earth several times. They gradually return to the earth over a period of a few years and consequently give appreciable doses to the world's population.

The dose from nuclides injected into the troposphere reaches a peak shortly after each weapon test.

The fall-out from atomic explosion also cause pollution in the atmosphere. The radioactive dust that falls to earth after atomic explosion is called radioactive fall-out. These materials mix and interact with natural particulate materials in the atmosphere and increase the amount of man-made air pollution.

The kind of radioactive fall-out depends upon the type of bomb. Fall-out from weapons differs from atomic waste materials in that the radio-nuclides are fused with non-silica, and dust and whatever happenes to be in the vicinity to form relatively insoluble particles.

The smaller particles adhere tightly to the leaves of the plants where they may not only produce radiation damage to plant's tissue but may be injected by grazing animals and then to their alimentary canal. The fall-out can enter the food chain directly at the hrbivore or primary consumere level.

In gereral, the total amount of radioactivity decreases with distance from a nuclear test. There are certain exceptions to this especially strontium-90 which reached a peak in wild animal populations 50 to 100 miles from 'ground zero' of the explosion. The amount of all-out received in an area is roughly proportional to the rainfall. the quality of fall-out radionuclides that enters food chains and eventually becomes transferred to man depends not only on the amount received from the air but also on the structure of the eco-system and the nature of its biogeochemical cycles .

In geeral a large proportion of all-out will enter food chains in nutrient poor environment. Fallout radionuclides have been and are being passed on to man through the food chain, although concentration in human tissues are not generally as high as those in sheep and deer. Man is somewhat protected by the position in the food chain and by food processing and cooking which removes some of the contaminants.

Radiations from Miscellaneous Sources

The cigarette smoker, as usual, reaps special harvests. Besides cancer-inducing tars, his lungs are getting some cancer-inducing radioactive lead (at least 40 millirems/ year for the pack-a-day smoker); it is roughly a third of what he cannot escape because of background radiation (which does not all go right to the lungs). There has been an estimated gram of radium, on the average, in the top 6 inches of each square mile of soil.

When radium decays it produces radon, another radioactive substance when is a chemically inert gas, which mingles with the atmosphere. The decay of radon produces a radioactive isotope of lead that comes out of the air and deposits on foliage like tobacco leaves. There have been also a number of other everyday sources of man-made radiation.

Older-type luminous watch faces using radium could give local dose of up to 2 mrem/h, but tritium (H^3) is used today and the doses are only a negligible portion of the man-made radiation. X-rays from black-and-white television sets are contributing another millirem per year and large doses have been occasionally reported from other sources –10 mrem/h from houses built with radioactive stone,100 mrem/h from houses built with radioactive stone, 100 mrem/h from bathtubs glazed with uranium pigments, etc.

Chernobyl Nuclear Mishap

On April 28th, 1986, the melting of a nuclear reactor and

its explosion had taken place at the Chernobyl nuclear power station, 80 miles north of Kiev in USSR. It is regarded the worst disaster in 32 years of commercial atomic power which has caused unspecified deaths, suffering and environmental and health damage. The dangerous radiation also affected neighbouring countries.

Around 2,000 people near the plant had been killed by causes which were ranging from the initial blast to lethal radiation and about ten thousand were evacuated from the endangered region. Radioactive gases and particles have spread over a vast section of the Soviet Union and water supplies for more than 6 million inhabitants of the Kiev area get threatened with contamination.

The most serious part of the nuclear accident was the radiation emitted from the reactor, which was then carried by winds to the neighbouring areas. In the first few hours of the disaster deadly form of iodine and cesium got released into the atmosphere along with other highly dangerous radioactive emissions.

By the end of the week a large invisible cloud of radiation covered Eastern Europe and the shores of the Mediterranean sea. The people who have been actually in trouble have been those right at the site and local residents who risked exposure to extreme doses of radiation that could give rise to cerebral haemorrhage, nausea, vomitting and death within hours.

The Most Dangerous Radioactive Pollutants

Strontium-90, cesium-137, and iodine-131 have been by far the most serious pollutants in the general environment from fallout and as atomic wastes. A radioactive isotope is regarded to belong the "serious pollutant" category if it satisfies the following criteria:

1. It is formed in a high yield in whatever nuclear reaction makes it (fission or fusion).

2. Its half-life has been in the range of 1-1000 years. (If much shorter, it decays rather quickly; if much longer, its activity is low).
3. It is easily taken up by a living organism—either man or something in the food chain of man.
4. It is not eliminated from important body organs (i.e. it has a long biological half-life).

Iodine-131 (half-life 8 days) has been a serious pollutant in spite of its low half-life because it has been produced in high yield by fission (both in weapons and reactors), it is efficiently taken by cows from grass and hay and put into milk, and one tiny gland, the thyroid of man and animals, very effectively concentrates it. Radioactive phosphate cesium and iodine-131 can also readily accumulate in plants and animals through natural food chains.

However, in food chains involving arthropods radioactive isotopes of potassium, sodium, and phosphorus accumulate, but isotopes of strontium and cobalt do not. A less common but locally important radioactive isotope classified as a serious pollutant has been plutonium-239. *Strontium 90* normally, is formed in radioactive fallout, has a half-life of 8.1 years and behaves like calcium in biogeochemical cycles.

Thus it gets absorbed by plants, ingested by animals and deposited in bone tissues close to blood forming tissue. Strontium-90 is able to concentrate in natural biological systems in following method: water → bottom sediments → aquatic plants → fresh-water clams → minnows and small fish → musk rats. It is experimentally proved that due to this food chain musk rats concentrate strontium-90, 35,000 times above the levels of the water in which they live. Grazing animals are able to concentrate strontium-90 by ingesting it through grass and forage, and it can then be passed on to humans through milk.

Its beta rays emerge with 0.55 megaelectron volt. In the environment it is concentrated in food chain of animals, being a bone-seeker. Strontium-90 ions are chemically similar to calcium ions, and at least on surfaces of bone tissue can replace some calcium ions. Although isotopes are able to accumulate in human tissues as well as those of plants and animals, it could not be established so far whether current levels of isotopes in human tissues represent serious health hazards to man.

Some medical scientists such as *Gofman* and *Tamplin* and *Sternglass*, however, feel that man's radiation exposure from artificial sources has been already sufficient to produce serious disease problems (leukemia and bone tumors), genetic damage, and infant mortality. Tritium is an isotope of hydrogen having a mass of 3 units. It is an emitter of relatively low energy beta rays and has a half-life of 12.36 years.

Its beta rays cannot penetrate intact skin, but tritium occurs as tritiated water (HTO) in which form it easily enters living organisms by food, drink, humid air, and even direct absorption through the skin. Tritiated water undergoes the same kinds of chemical reactions as ordinary water, providing a route for the distribution of tritium throughout the entire body. Some of it inevitably becomes incorporated into the basic genetic material, DNA, where it may cause molecular change leadings to tumors, cancer, or fetal abnormalities.

Most but not all of the studies involving exposure of experimental animals to ingested tritium, either as tritiated water or as tritiated DNA (the chemical of genes), have shown that tritium has no unique and special powers of damage compared with X-rays or gamma rays.

A dose from tritium has the same radiobiological and radiation protection meaning as the same dose, with the same rate pattern, of X-rays or gamma rays. Therefore, tritium can cause cancer, and this effect has been demonstrated in laboratory mice exposed to the isotope. By the year 2000, assuming that nuclear power enjoys a heavy growth, the exposure of the general population to tritium released from nuclear power plants will average not more than 0.001 millirem/year.

This may be compared with the current contribution of tritium to our backgroung radiation of 0.06 millirem/year. Consider also the roughly 120-140 millirem/year from watching color television, or the several millirem from exposure in a long jet flight or a high-altitude sking vacation. Average exposures, however, tend to hide what might be serious local effects near power plants or atomic fuel reprocessing plants. The maximum permissible concentration of tritium in water is 0.0003 microcurie per cubic centimeter.

If the seven nuclear power plants planned for the shores of Lake Michigan are built, they will discharge tritium into the lake until by the year 2000 its concentration will be about 0.2% of this maximum. Tritium is produced in nuclear reactors in a number of ways. Some comes from relatively infrequently occurring fission events—one tritium atom in every 10,000 fissions. Coolant additives (e.g, boric acid or lithium hydroxide) capture escaping neutrons and undergo subsequent nuclear reactions that release tritium. Light water (i.e, ordinary water) contains traces of deuterium, another isotope of hydrogen, which can be changed to tritium by neutron capture. (That is why tritium releases from heavy water reactors used in Canada and several other countries are much higher than from light water reactors in the United States.)

Compounds of lithium and boron are present in control roads, where they interact with neutrons to produce tritium. Plutonium is a man-made, silvery, metallic element (atomic number 94). It forms as a bay-product from uranium-238 in burner reactors of nuclear power plants, and it would be continuously made in proposed breeder reactors.

Plutonium 239 is the most important of several plutonium isotopes. Thus far its chief use has been to make warheads of atomic bombs or the triggering devices of hydrogen bombs. The importance of plutonium-239 in civilian nuclear power is growing (see Breeder Reactor; Atomic Energy).

In the Nuclear Energy report of Project Independence (November 1974). the Federal Energy Administration assumed that nuclear power plants would obtain plutonium fuel from the reprocessing of spent uranium fuels starting soon after 1977. Since a number of dangers exits, this projection might be optimistic.

When plutonium-239 enters the environment from weapons tests or accidental releases from reprocessing plants, it appears as plutonium oxide, a very fine dust that is insoluble in water Plutonium-239 emits powerful alpha and gamma rays and has a half-life of 24,360 years. Its alpha rays are about 10 times more harmful in lung tissue than X-rays or gamma rays. Plutonium-239 is probably the most poisonous substance known. Inhaled into lungs, it causes lung cancer. If it enters the bloodstream, it causes bone cancer.

The greatest danger of plutonium-239 lies in its potential as a fuel in atomic bombs. Every nuclear power plant produces this material as its uranium fuel 'burns'. Plutonium-239 can be separated from spent uranium fuel-about 25 kilograms of Plutonium-239 from one year of operation of a 1000-megawatt power plant.

A simple implosion atomic bomb requires 4-8 kilograms of plutonium, and atomic bomb. When nuclear reactors are sold to countries wanting to join the "nuclear club", the countries may be expected to develop their own atomic bombs in 6-10 years. As more and more countries develop their own bombs, the possibilities that one will attempt nuclear blackmail against others increase.

The theft of plutonium between the time it is produced in a reprocessing plant and the time it is loaded into a nuclear reactor is another cause of grave concern in government and environmentalist circles. With only modest scientific expertise, a terrorist group could make a small atomic bomb from stolen plutonium.

Radiation Protection Standards

Soon after the discovery of X-rays and radio activity a sizable number of workers started using ionizing radiations. Their number increased manifold after the start of nuclear energy programmes in different countries. Hundreds of thousands of works have been associated with this programme directly while millions of people in a number of countries have been affected in some way or the other by nuclear radiations.

Right from the early days of discovery of ionizing radiations the deleterious effects of over exposures to radiations had been observed. Therefore, a need to prescribe some guidelines for the use of radiations was felt. It was only in 1928 that a unit of measuring the quantity of radiation exposure, the roentgen (R) was established. There were obvious difficulties in formulating standards of radiation dose due to lack of information on the correlation between dose and the injury it produced in an organisms.

For instance, it is very difficult to determine the limit of radiation below which no significant injury will result. No direct expermients can be conducted on human beings. Extrapolation of findings on animals is bound to be in error due to differences among animals of different species. Inspite of these handicaps a large amount of information on the radiation effects has become available from the use of X-rays and radium sources in radiology and cancer therapy, from victims of atomic blasts accidents at nuclear energy installations, and general population exposed to atomspheric tests conducted by the major and less major powers.

Therefore, the radiologists and health physicists have formulated certain broad proposals which have been accepted as guidelines for radiation protection. Some of the principles underlying these guidelines pertain to the recovery from the effects of small radiation doses. The body cells are replaced quickly or slowly depending on the nature of the tissue.

Consequently a minor somatic injury is cured with time. However, the genetic effects are cumulative and there is no recovery from them. Secondly all living beings are exposed to cosmic rays and other ionizing radiations continuously. Radiation dose received from cosmic rays is 400 to 500μ Sv y^{-1}, another 900μ Sv y^{-1} is received from terrestrial sources like residences, soil, clothes, etc., while another 200 to 300μ Sv y^{-1} arises from sources inside the body namely, water, found, air, ^{40}K, ^{226}Ra, i^4C, etc., so that on an average a dose of 1500μ to 1700μ Sv is received annually from natural sources and the human body has adjusted itself to this without any apparent damage.

On the international level the only widely recognized authority for setting maximum permissible exposures to ionizing radiations is the International Commission on Radiological Protection (ICRP), founded in 1928. It was chiefly concerned with the safest possible uses of X-rays and radium consistent with the human benefits these were known to offer. From the agency's inception, ICRP policy has been to permit no necessary exposure. The problem has been in defining what is necessary.

When long-term risks are impossible to assess by short-term experiments. How can we know when benefits outweigh risks? The benefits seen most obvious for medical uses, and therefore, the ICRP has allowed exposures in this area to be 30 - 100 times the levels deemed acceptable for the average person in the population at large.

Giving the patient such exposures, of course, means extra risks to the medical personnel who handle this work or, in the case of radium, to people who mine and process radium ores. Thus the category of occupational exposure limits came into being. Since its inception, the ICRP has been the one internationally recognized body which is responsible for recommending values of maximum permissible exposure to ionizing radiation. Most of the national legislation in many countries relating to exposure to radiation has been based on the recommendations of the ICRP. The early recommendations of ICRP were concerned with protection against X-rays and radium. Some of the earliest recommendations were dealing with the length of time that a worker should get engaged on radiation work; these were:

(a) Not more than 7 h/day.
(b) Not more than 5 days/week.
(c) Not less than 1 month's holiday per year.
(d) Off days to be spent as much as possible out of doors.

The maximum permissible doses had been defined very loosely.

In 1950, the ICRP tried to extend its scope to deal with the many new problems which had resulted from the introduction of nuclear reactors. All of these ranges have resulted in a decrease in the occupational exposure values. This continuous decrease in the maximum permissible exposure level has not been attributed to any positive evidence of damage to persons working within the earlier permissible dose levels, but rather to an increasing awareness of the uncertainties in much of the available experimental data.

Table 7

Values of Maximum Permissible Occupational Exposure

Dose rate	*Date recommended*	*Comments*
0.1 of an erythema	1925	Proposed by A. dose per year Mutscheller and R.M.Sievert. This corresponds to an exposure of ~30 R/year from 100 kV X-rays or ~70 R/year from 200 kV X-rays.
0.2 R/day or 1 R per working week	1934	Recommended by I.C.R.P.
15 rem/year, or approximately 0.3 rem/week	1950	Recommended by I.C.R.P.
5 rem/year, or approximately 0.1 rem/week	1956	Recommended by I.C.R.P.

Over the years many different expressions are being used to describe acceptable levels of exposure to radiation. The most important of these have been tolerance dose and maximum permissible dose.

The term tolerance dose was having the unfortunate connotation that it seemed to imply a threshold dose below which no radiation damage would take place. On the basis of presently available evidence there has been no proof of the existence of a threshold dose for certain types of somatic damge.

Consequently, the term tolerance dose has been not employed now-a-days but has been replaced by the term maximum permissible dose (MPO). In 1958, the I.C.R.P. postulated the following definition of MPD: 'The permissible dose for an individual has been that dose, accumulated over a long period of time of resulting from a single exposure, which in the light of present knowledge carries a negligible probability of serve somatic or genetic injuries; furthermore, it has been such a dose that any effects that ensue more frequently get limited to those of a minor nature that would not be considered unacceptable by the exposed individual and by competent medical authorities.

Any serve somatic injuries (e.g., leukemia) resulting from expsure of individuals to the permissible dose would get limited to an exceedingly small fraction of the exposed group; effects such as shortening of life span, which might be expected to take place more frequently, would be very slight and would likely be the hidden by normal biological variations. Thus permissible doses can be expected to produce effects that could be detected only by statistical methods applied to large groups.'

Balancing the Risk

The problem in setting MPD values has been to balance the expected benefits to get derived from the use of radiation and atomic energy against the risk from radiation exposure.

If there is the absence of any definite evidence of a threshold dose for certain types of somatic damage, it must be regarded that there is some risk associated with a dose of any size. It may be very difficult to define the proper balance between risk and expected benefit and it may vary considerably from country to country, depending on social and economic factors.

The responsibility for setting levels ensuring a correct balance between risk and benefit has been very considerable. If we set MPD's has been at too high a level, an unacceptable degree of injury may get sustained. On the other hand, if the MPD's are unncessarily low, many, of the benefits of radiation and atomic energy would get lost.

Recommendations of I.C.R.P.

The Commission's recommendations have been attempting to limit somatic effects in individuals and hereditary effects in the population as whole. It therefore becomes necessary to consider not only dose to individuals but also the average dose to the population. We know that some organs are more sensitive than others to the effects of radiation. Therefore, different limits have been specified for various organs.

Table 8

Summary of Does Limits for Individuals

Organ	*MPD for adults evposed occupationally*		*Members of the public*
	Rem/year	*Rem/13 weeks*	*Rem/year*
Gonads, red bone marrow	5	3	0.5
Skin, bone, thyroid	30	15	3.0
Hands, forearms, feet and ankles	75	38	7.5
Other single organs	15	8	1.5

The following points must be stressed in applying these does limits :

(i) All unnecessary dose must be avoided.

(ii) Adults exposed in the course of their work who might exceed 3/10ths of the MPD should be subject to medical supervision and personal monitoring. Such persons are often called Classified Radiation Workers.

(iii) The permissible quarterly (13 week) doses for occupational exposure have been based on one-half of the annual dose, but are rounded up. The object of the 13 week MPD's has been to limit to some extent the rate of accumulation of dose. For example, if only annual limits were specified it would be possible for a worker to receive 5 rem (whole body) on the 31 December and another 5 rem on 1 January, providing a total of 10 rem in 2 days. This is obviously undesirable and is only permitted in very special circumstances (see below).

(iv) In special circumstances it may be justifiable to allow the quarterly quota to get repeated in successive quarters provided that the total body dose accumulated dose not exceed 5 (N - 18) rem, where N denotes the age of the person in year. Thus it has been permissible under special circumstances for a person to receive in a single years dose of 12 rem to the gonads and rad bone marrow always provided that the D = 5(N - 18) rule is obeyed.

(v) Women of reproductive capacity should not get exposed at the increased whole body rate of 3 rem/ 13 weeks, but should get limited to 12.5 rem/13 weeks.

(vi) The suggested limits for members of the public should be in all cases a factor of 1/10 of the equivalent limits for adults exposed in the course of their work.

(vii) In the case of uniform whole-body irradiation, the gonads and rad bone marrow are regarded to be the critical organs, so for uniform irradiation the whole body dose limits have been the same as those for the gonads and red bone-marrow.

Planned Special Exposures

In very special circumstances, it has been permissible to receive doses of up to twice the annual limit provided that the 5(N-18) rem limit does not get exceeded and the quarterly limit does not get exceeded in the last twelve months. Planned special exposures should not be allowed to a worker who has previously received abnormal exposures in excess of five times the permissible annual dose, or to women of reproductive age. It is generally felt that planned special exposures should only be allowed under very unusual circumstances where no alternative solution has been possible.

Abnormal Exposures in Emergencies or Accidents

It is obvious that it is not possible to specify dose limit for accidental exposures. However, I.C.R.P. have recognized the need to protect the livelihood of a worker involved in such an accident.

They recommend that no special limitations should be kept on his future radiation dose following a 'one-in-a-lifetime' accidental exposure of less than 25 rem. It is important to remember that when a worker has used up his once-in-a-life-time special exposure allowance he severely gets limited as far as future radiation work goes. After such an exposure, he must be used on work which has an extremely low probability of leading to another high dose.

In an emergency, male valunteers may get large doses for the purposes of rescue or of saving valuable installations. It has been difficult to specify limits to cover such an event as each situation will be unique. It has been often recommended that a whole body dose limit to 10 rem should be applied if saving of life is not involved. When rescue operations are to be carried out it may not be possible to work to specified dose limits.

Each situation has to be assessed by the competent person or persons, and a decision reached on the basis of this assessment. It has been probable that whole-body doses of up to 100 rem could be acceptable in the saving of a life. If the operation would need doses much in excess of this level, then the risks and possible result of the operation would be judged very carefully. One important point would be the accuracy of the information regarding the likely dose-rates in the accident area and a second, the condition of the casualties and their likelihood of survival. For example, if the dose-rate estimates were low by a factor of two, a rescuer might unwittingly receive a sufficiently large dose to produce severe early somatic effects.

The main contributors to the exposure of the general population have been natural radiation, medical uses of radiation, the increasing use of radioactive substances in industry and the release of readioactive material to the environment. Two methods have been employed to keep this exposure as low as possible, such as the reduction of individual doses and the limitation of the number of persons exposed. The genetic dose to the population refers to the average received by each person from conception to the mean age of child bearing, which has been taken as 30 years.

2 Radiation Pollution

Thermoluminescent (TL) Dosimeters

Thermo-luminescent (TL) dosimeters have been currently among the most reliable and accurate instrument which are useful in a very wide range of does from 1 μGy (0.1 m rad) to 10 Gy (1 K rad). A crystalline material such as calcium fluoride (CaF^2) or lithium floride (LiF) or a similar compound in powder form is incorporated on a plastic material like teflon.

Irradiation of the material produces electron trapping at the crystal lattice sites at both shallow and deep traps. On heating the material the electrons are released. They produce luminescence on de-excitation. A photo-multiplier measures the amount of light produced. The light output has been proportional to the received dose.

A typical TL Dosimeter is in the form of a pen (or ring) in which the sensitive material (LiF crystal rod) and filament are placed in a small vacuum tube. The filament is heated when the tube is plugged in a voltage supply. A number of manufacturers market complete TLD systems for the preparation of TLD badges and arrangement for their processing after their exposure to radiation. Finger monitors in the form of rings of TLD material for monitoring the additional dose received by the fingers and hands are also available.

To charge a pocket chamber or desimeter, a charger containing a battery or rectified power supply is used to apply voltage to the chamber electrodes. The voltage is adjusted until the shadow of the fiber falls at 0 on an illuminated scale. After exposure to radiation, a pocket chamber must be reinserted into the charger (termed the charger-reader) to determine the deflection of the fibre upscale.

The pocket diosimeter contains its own scale, and the deflection of the fibre upscale may be tetermined by holding the dosimeter toward light. When both neutrons and gamma rays are to be measured the TL dosimeters with discs of LiF prepared from separated ^{6}Li and ^{7}Li isotopes are used. ^{6}Li has a high cross section for ^{6}Li (n, α)^{3}H reaction with thermal neutrons whereas ^{7}Li is relatively insensitive to neutrons.

Both the isotopes are however equally sensitive to gamma rays. Filters are used for discriminating between beta and gamma radiations. By using lead filters the response of the device becomes energy independent for X-and gamma rays in an energy range of 20 keV to 20MeV. The TL dosimeters have certain memory component. When they are heated for a short time only the shallow traps are emptied of electrons to produce luminescence.

On further heating to a higher temperature the electron at deep traps produce light output. In this manner it is possible to make a second measurement of the same dose at a later time. A TL dosimeter is used over and over again. This is done first by erasing the memory of an earlier dose by heating the dosimeter to a high temperature. Then it is ready for making fresh measurement of dose.

Thermo-current (TC) Dosimeters

An improvement-cum simplification over the thermoluminescent (TL) dosimeters has been achieved by developing thermo-current dosimeters. The principle of this device has been similar to that of TL dosimeters. Radiation produced electrons are trapped at impurity levels in a thin crystal of synthetic saphire which is 10 mm in diameter and 1 mm or less in thickness in a typical device. The trapped electrons get liberated on heating to produce current in a circuit formed by two electrodes deposited on the crystal surfaces. Then the peak value of current or the integrated current is a measure of absorbed radiation dose.

This device eliminates the use of photo-multiplier tube. Further it has been more convenient to measure small currents than feeble light output. Therefore the TC dosimeters are likely to replace the TL dosimeters. The latter have already replaced the film badge dosimeters to a large extent.

Film Dosimeters

Photographic emulsions have played vital role as integrating detectors for directly and indirectly ionizing radiations. Photographic films have long been in use in dosimetry, especially with X-rays. Under controlled processing and good densitometric measurements film dosimeters have given generally reliable results.

An exposed film and a reference film are processed together. The radiation exposure time is usually one week but it can be less if an accidental over-exposure has taken place. Measurements of the densities of the blackenings of exposed and reference films and their comparison with suitably calibrated standards with known radiation doses give the radiation dose to which the worker has beeen exposed.

The normal range of doses measured with these devices has been 500μ Gy to 200μ Gy (0.5 to 20 rads).

These limits can be extended by using different types of films. Nuclear emulsions are used for measurement of dose from fast netrins and high energy particles produced by accelerating machines. Special boron loade films are used for the thermal neutrons. Although the film badge dosimeters have been in use for several decades and are still being used in several establishments as routime personnel monitors, they are being replaced by TL and TC dosimeters.

Direct Reading Ionization Chambers

The amount through which the anode potential has dropped in a given times is a measure of the gas ionization and hence the exposure from incident radiation during that interval. In a direct reading instrument, like a pocket type or a pen type ion chamber, the movement of metallized quartz fibre across a calibrated scale indicates the total dose received. For more accurate measurement auxiliary circuitry is used with other versions.

These instruments are generally used for the measurement of daily dose received by the workers in contrast with the film badges which measure weekly doses. The capacitor in charged to a full voltage at the beginning. As the exposure builds up the capacitor gets discharged. The reading on the scale, calibrated in terms of roentgen, rads or Gy gives the dose received at the end of a working day or an operation.

These instruments have ranges from 200 Gy to 10 Gy (20 mrad to 1000 rad). Usually, instruments with 1000 to 2000μ Gy (100 to 200 mrad) full scale ranges are used in the moral laboratory work. Chambers lined with boron are used in work with slow neutrons. *Auritsen Electroscope:* This device is similar to the pocket dosimeter, except that it is bulkier (the chamber is 2-3 in, in diameter and about 3 in, long) and is used primarily for measurement of radiation from radioactive samples. In the instrument, the deflection of the quartz fibre is viewed through the microscope at one end of the chamber. The Landsverk electroscope or analysis unit is somewhat șimilar to the Lauritsen electroscope, except that the sample may be inserted and measure within the chamber.

The size of the pulse produced by single ionising particle may be easily calculated as follows. About 30 eV are needed for an alpha to form an ion pair in a gas. A high energy alpha of 60 MeV will produce about 2 x 0^5 ion pairs. Now one electron carries a charge of 1.6 x 10^{19}C, so that the arrival of this seemingly large number of electrons at the anode will generate a charge of only 3.2 x 10^{-14}C.

It is therefore apparent that ion currents, even with large fluxes of energetic particles, are very small. These will be even smaller with the smaller fluxes of lower energy particles that are more commonly encountered, as with the beta-emitting isotopes used in tracfer experiments. Such small currents are difficult to measure directly. In instruments other than the electroscope they are amplified before measurement.

Phosphate Glass Dosimeters

Dosimeters described above are useful for personnel monitoring. They have a narrow and limited ranges of operation. In problems connected with high doses of radiation like those in atomic blasts; reactor or accelerator flux densities; irradiation of foods, medical supplies etc., glass dosimeters are handy.

A glass dosimeter has been a small strip of a special type of glass, a few cubic cms in size, containing small amounts (5 to 10 per cent) of a phosphate with silver impurities. Usually silver phosphate in polymer form is used. The strip is coated with a thin layer of an opaque material to protect it from visible light. Ionizing radiation produces electrons which are trapped at impurity centres and stay there permanently until released by ultra-violet radiation. Then give rise to liminescene which is measured with a photo-multiplier-recorder assembly.

The amount of emitted light is a measure of dose received at the dosimeter. This devices could be used for recording doses from a few Gy to several k Gy which are used in irradiation units for preservation of foods and sterilization of medical equipment.

Chemical Dosimeters

A dye is used as an indicator of the pH-value of solution which depends on the amount of HCI formed. In one version choloroform in aqueous solution and the indicator dye are contained in a sealed glass tube. Irradiation of solution produced a change in the colour of the dye. The dose is then obtained from the extent of change of colour. Fricke dosimeter, which uses ferrous solution, is based on the oxidation of ferrous ions (Fe^{2+}) into fetric ions (Fe^{3+}) under irradiation. This dosimeter is used in X-ray dosimetry. An aqueous solution of ferrous ammonium sulphate (0.001 M), sodium chloride (0.001M), and sulphuric acid (0.4M) is usually employed. Radiation changes Fe^{2+} ions into Fe^{3+} ions.

The concentration of the latter is then a measure of radiation dose. The amount of Fe^{3+} ions is determined by spectrophotometric method. This dosimeter has been used to measure doses as high as 10^7 to 10^8 Gy per sec. Its response has been independent of the dose rate upto a dose of 10^5 to 10^6 Gy per sec but depends on the presence of absence or oxygen. Ceric sulphate dosimeter gets on the reduction of ceric ions (Ce^{4+}) into cerous ions (Ce^{3+})

under irradiation. Its response is independent of the presence or absence of oxygen.

The dose is measured from the yield of Ce^{3+} ions. This is obtained from the difference in ceric ion concentrations before and after irradiation. Optical and analytical methods are used for measurement of Ced^{4+} ion concentration. This dosimeter shows a linear response in a region of 10^3 to 10^5 Gy. Doses upto 10^6 Gy could be measured with this type of dosimeter.

Calorimetric Methods

Absorption of radiation by a medium raises its temperature. If the changes in temperature could be neasured accurately then it becomespossible to determine the amount of energy deposited in the medium by the incident radiation. This forms the basis for calorimetric methods. For an ionizing radiation imparting 10 Gy of dose to a material the rise in temperature is only a few millidegrees.

The success of this method therefore depends on the accuracy with which small changes in temperature could be measured. The importance of calorimetric method lies in its being an absolute method against which the other indirect dosimetric measurements may be checked.For an experiment, a sample of water or graphite is kept in a thermostatically controlled enclosure, insulated thermally from the environment. It is usual to follow the small temperature drifts before and after the measurement of rise in temperature due to a beam of incident radiation. As the minor drifts in temperature can not get eliminated completely therefore an extrapolation procedure is used to determine the rise in temperature.

A precision of about 1 per cent in the measurement of temperature difference is unually obtained. The calorimetric method gives the rate of energy production in the medium. This gives the activity of the source if the average energy of alpha or beta particles emitted by the source is known. Energy production rates of the order of a few micro-joules per sec have been measured with an accuracy of 1 per cent. It becomes necessary to ensure that radio-chemical reactions do not modify the changes in temperature. This is usually carried out by pre-irradiation of the system. In that case any chances of further chemical changes could get reduced.

Radiation Monitors

A variety of instruments including ionization chambers, GM-counters, and scientillation counters with rate-meter dispkays, an available commercially. They are usually known as survey meters or contamination monitors and calibrated in terms of counts per sec or counts per min or m Gy (or rad) per hour. Both mains operated and portable varieties are commonly available for the laboratories. A survey meter with ionization chamber detector, dc amplifier, and calibrated meter has ranges from 200 μGy per hour to 30 m Gy per hour.

A removable cap, which works both as a protection for thin window as well as a beta particle shield, enables the monitoring of both beta-gamma and gamma activities. Aportable Gm-counter Survey Meter with cylindrical type of countries is more sensitive instrument with ranges from 2 μGy h^{-1} to 200 μGy h^{-1} (0.2 m rad h^{-1} to 200 m rad h^{-1}). Here again a thin shield which covers the tube allows the instrument to detect only gamma rays above 100 keV energy while with the shield removed it is sensitive to both beta (above 0.2 MeV) and gamma rays. Thin window GM tubes have been very useful probes for beta particle monitoring. Soft beta rays from ^{14}C and ^{38}S need extremely thin windows. A Scintillomwter using Nal (TI) Photomultiplier rate-meter system has still better sensitivity for gamma rays. It measures radiations does as low as 0.01. Gy per hour.

Thin NaI (TI)crystal (1 mm thick) scintillation probes are available for use in 5 to 50 leV region. They are particularly useful for low energy photons from ^{325}I. Contamination monitors with warning systems also form part of radiation protection equipment. An instrument fitted with a small speaker or some other relay operated warning system becomes operative when the radiation dose exceeds a certain limiting value.

Levels of radio-activity in air are monitored with radio-active gas monitors. Air is withdrawn through an ionization chamber. The level of radio-activity due to alpha or beta emitters in gaseous forms has been then indicated by audible or visual displays. In an another version a small vacuum cleaner is used to draw out air and dust into a filter paper. The latter could be then monitored for its radio active content.

Nuclear Energy and the Environment

The first stage in the nuclear industry is the uranuim mine. The ore is pulverized and subjected to a leaching process to separate the uranium. The waste from this separating process, known as tailing, contains the subsidiary products of the natural radioactive series and is deposited on dumps.

In the United States alone, the volume of these tailings already amounts to some 100 million tons. Rainwater washes the soluble nuclides into the ground, into groundwater and into surface waters, whence it is absorbed by organisms in the soil and water. By the extraction of uranium from the mines, the natural radioactive substances that normally lie deep down in the earth's crust are brought to the surface and released into the biosphere. The uranium extracted from the mines then passes into the factories for preparing fuel elements, where the fissionable uranium-235 isotope recovered from nature in small quantities is enriched. In this highly energy- consuming enrichment process further radioactive waste is formed and given off into the environment. Once the uranium has been sufficiently enriched it is fed, in the form of small tables, into fuel elements for nuclear power plants. In the nuclear power plant the real process of nuclear fission takes place, whereby a large number of mostly radioactive fission products are formed from the uranium-235.

These partly gaseous radioisotopes escape to some extent in the nuclear power plant and are given off into the air and water below the legally permitted limits for radioactivity. After about a year the fuel elements are exchanged and, after a period of intermediate storage in the nuclear power plant to permit decay of the short-period isotopes, are transported to a reprocessing plant. For reconditioning purposes the fuel cells are split by remote control, the individual roads are cut into short sections, and the uranium oxide is washed out of the sheaths with nitric acid.

By this disintegration of the fuel elements the radioactive inert gases cannot be formed by nuclear fission are released. As inert gases cannot be combined by a chemical process, they are emitted from the reprocessing plant through a chimney into the atmosphere. Other radionuclides released in gaseous and liquid form in this process are kept out of the waste air and water of the plant means of filters. But as these filters are not 100 per cent absorbent, a certain amount of the radioisotopes escapes into the environment.

Overall the emission of radioactivity by reprocessing plant, related to the corresponding fuel equivalent, is higher by a factor of 100-200 than in nuclear power plants. The uranyl nitrate solution produced in the dissociation of the fuel elements is subjected to a multistage extraction process, by means of which the uraniuim the plutonium and the fission products are separated from one another.

Biological Effects of Ionizing Radiation

The magnitude of their injurious effect has been in a range that might be said figuratively to represent a "quantum jump" in comparison to ordinary organic chemical poisons. Radium has been 25000 times more lethal than arsenic.

The biological importance of radiation were known when in 1895 Wilhelm Roentgen kept his hand between the newly discovered x-ray tube and a flourescent screen and noted that the bones cast a deeper shadow than the ffesh. The following year, the French physicist Antoine Becquerel discovered natural radio activity in a uranium compound, a discovery for which he shared the Nobel Prize with the Curies in 1903.

An accident with one of the earlest known radioactive substances gave Bequerel the first indication of the biological potency of the newly found radiation when he got burnt himself with a vial of radium that hc was carrying in his pocket. The most tragic early evidence of the potency of radiation had been death of Marie Curie, although the lesson seemed to go unheeded for many years. Marie Curie with the help of her scientist husband. Pierre Curie, found that uranium are had been having a substance as yet unidentified that could be several hundred times as radioactive as the uranium with which she was working.

They were successful in isolating from the ore a radioactive element which they named polomium after Maries native country Poland. But polonium accounted for only a fraction of the radioactivity, and in 1898 they identified radium, an intensely radioactive substance.

Pierre, upon learning of Bacqerel's experience with radium, carried out an experiment that deliberately produced a burn on his own arm. Marie Curie became famous. Although she was denied membership in the snobbish French Academy by one vote because she was a woman, she was able to share the 1903 Nobel Prize with her husband and Bacquerel. In 1911, she again received the Nobel Prize, this time for the discovery of two new elements. Marie Curie died of leukemia, a cancer of the formative white blood cells of the body, almost certainly brought on by the high levels of radiation to which she was exposed during her work.

Irene Joliot-Curie, the scientist daughter of the husband and wife team, also died, of leukemia apparently from radiation exposure. When x-rays were discovered, their usefulness in medical diagnosis and other applications was quickly recognized. However, the dangerous nature of x-radiation and its potential for serious, even fatal, injury were largely unknown. Indiscriminate use of x-ray machines brought about numerous cases of injury and illness.

Miners suffered toxic effects from radium and thorium and the careless use of radioactive materials, most notably radium, made serious ailments and premature deaths among workers. The potency of x-rays and radionuclides has been so great that the enormously expanded use of ionizing radiation in its many forms in research, medicine and industry is potentially hazardous people working in areas of high radiation and in the environment at large.

The interaction of ionizing radiation with the human body, arising either form external sources outside the body or from internal contamination of the body by radioactive substances, give rise to biological effects which may later show up a clinical symptoms. The nature and severity of these symptoms and the rate at which they appear depends on the amount of radiation absorbed and the rate at which it has been received.

Radiations injuries could be divided into two classes, somatic effects in which the damage appears in the irradiated person himself and genetic effects which arise only in the offspring of the irradiated person as a result of radiation damage to germ cells in the reproductive organs- the gonads.

The radiation released upon the disintegration of radioactive atomic nuclei takes the form of particulate or wave radiation of very high energy. When this radiation impinges on living cells, various interactions can result. The most important of these is the ionization of atoms. In the electron sheath of an atom, a certain number of negatively charged electrons are present which correspond to a like number of positively charged protons in the atomic nucleus.

Consequently an atom is normally electrically neutral in its outward effects. If radioactivity, such as a negatively charged electron of beta particle radiation, bombards a neutral atom, it may be that owing to the high energy of the radiation one or more electrons will be ejected from the electron sheath of the atom. The atom then becomes an ion a positively charged (electrically) atomic residue.

Owing to their electrical attractive power, newly formed ions have a capacity for strong chemical reaction. If this takes place in dead matter, in the air for instance, it is of no importance. An ionized nitrogen atom, for example, can combine with ionized oxygen atoms; and ionized oxygen atoms can combine three at a time to form one ozone molecule. But if this process of ionization of atoms takes place in living tissue, in certain circumstance a variety of damage may result. It is now known that the plan for the constitution of a living organisms is contained in a long filamentary structure, in which, on the basis of a percise arrangement of small molecules, information regarding heritable characterists is contained.

This is deoxyribonucleic acid (DNA) and it consists of a basic structure of sugar and phosphate molecules to which four different basic molecules are attached in a particular sequence. This sequence is what is known as the genetic code, which represents the heritable information. In the case of each cell division the DNA is also divided and restructured on identical lines, so that the heritable information is fully retained in each cell in the case both of the restructuring of body cells.

If radioactive rays now impinge on a DNA molecule, there is a danger that, with the ionization of atoms and the chemical reaction ensuring therefrom, a variation may occur in the basic arrangement and so in the heritable information. Such a variation in the heritable information is known as a mutation. Almost all the mutations are negative, it is estimated that over 99.99% of the mutations change the organism's characteristics negatively. If the mutation occurs in germ cells that is in male sperm cells or in female egg cells-this can result, when a new organisms is formed, in severe damage, such as deformities, stillbirths, enzyme defects amd other metabolic disorders.

In body cells, otherwise known as somatic cells, a mutation can cause cancerous degeneration. Ithas to be assumed that not tolerance limits exist as regards this type of damage that is to say, that the very smallest quantity of radioactivity is capable of causing such damage. High radiation exposure can touch off a chain reaction via ionization of atoms, chemical reactions, toxic molecules reaction and poisoning of the entire organism, leading to immediate danger in the form of fever, nausea, vomiting, loss of hair and death from radiation.

Types of Biological Effects

Ionizing radiation is different from ordinary environmental contaminants in several ways. One of the most important is that some of the radioactive elements continue to emit radiation for extremely long periods of time—in some cases for thousands of years comprising a special kind of hazard to man and to the environment. The increasing quantity of radioactive waste products poses problems that are unique and require greater knowledge than is now available concerning both physical and biological aspects of the long-term effects.

The Interaction of Radiation with Cells

The basic difference between nuclear radiations and the more commonly encountered radiations such as heat and light is that the former have sufficient energy to cause ionization. In water, of which cells are largely composed, ionization can lead to molecular changes and to the formation to the chromosome material. The damage takes the form of changes in the construction and the function of the cell. In the human body, these changes could be able to manifest themselves as clinical symptoms such as radiation sickness, cataracts or, in the longer term, cancer.

The process leading to radiation damage could be conveniently divided into four stages as follows :

(i) *The initial physical stage*, lasting only a minute fraction ($\sim 10^{-16}$) of a second in which energy gets deposited in the cell and causes ionization. In water the process may be written as follows :

$$H_2O \xrightarrow{\text{radiation}} H_2O^+ + e^-$$

where H_2O^+ is the positive ion and e^- the negative ion.

(ii) *The physico- chemical stage,* lasting about 10^{-6} seconds in which the ion interact with other water molecules resulting in a number of new products. For example, the positive ion dissociates,

$$H_2O^+ \longrightarrow H^+ + OH$$

the nagative ion, that the electron, attaches to a neutral water molecules which then dissociates

$$H^2O + e^- \longrightarrow H_2O^-$$

$$H_2O^- \longrightarrow H+OH^-$$

Thus, the products of the reactions have been H^- OH^- H and OH. The two former ions,which are present to quite a larger extent in ordinary water, take no part in subsquent reactions. The other two products H and OH, are called free radicals, that is they have an unpaired electron and have been chemically highly reactive. Another reaction product has been hydrogen peroxide H_2O_2 which is a strong oxidizing agent.

$$OH + OH \longrightarrow H_2O_2$$

(iii) *The chemical stage,* lasting a few seconds in which the reaction products interest with the important organic molecules of the cell. The free radicals and oxidizing agents may attack the complex molecules which form the chromosomes. They may, for example, attach themselves to a molecules or cause links in long chain molecules to be broken.

(iv) *The biological stage* in which the time scàle varies from tens of minutes to tens of years depending on the particular symptoms. The chemical changes described above can affect an individual cell in a number of ways. For example, they may result in :

(a) the early death of the cell,

(b) the prevention or delay of cell division, or

(c) a permanent modification which is passed on to daughter cells.

The effects of radiation on the human body have been the result of damage to the individual cells. These effects may be conveniently divided into two classes, namely somatic and genetic. The somatic effects arise from damage to the ordinary cells of the body and affect only the irradiated person.

The genetic effects, on the other hand, are due to damage to the cells in the reproductive organs, the gonads. The important difference is that, in this case, the damage may be passed on to the person's children and subsequently to later generations.

When ionizing radiation happen to pass through body tissues, it is able to split the molecules of the tissues into ions or free radicals, as in the examples of water molecules. These fragments may later combine to form new chemical compounds (such as)H_2 from H+H or H_2O_2 (hydrogen peroxide) from OH+OH) or free radicals (such as HO_2 from H H+O_2).

Ionizing radiation can thus split molecules into useless or reactive fragments and allow the formation of other reactive compounds. These effects can be classified into two groups:

1. *Direct effects.* Fragmentation of biologically important molecules, such as DNA molecules in the cell nucleus.

2. *Indirect effects.* Fragmentation of biologically less vital molecules (such as water) with the formation or reactive ions or free radicals that can later affect more important molecules and impair their usefulness.

The damage caused by readiation has been found to depend on the amount of energy that gets lost by the radiation and made available for the ionization for matter.

The amount of energy lost by the radiation per unit distance travelled (which is not necessarily in a straight line-beta rays, for example, are easily detected by matter) is termed as the radiation increases as the energy of the radiation decreases, in part because its speed will be less and the traversal of a unit distance will require a longer period of time. A particle entering tissue with high energy will have low LET initially but as its energy decreases, its LET increases.

Such a particle could be used in treating a deep-seated tumor if the particle's range is just right for it to come to a stop near the tumor. The site in a living cell most vulmerable to ionizing radiation has been its nucleus, where most of its genetic material (DNA) has been located. If ionizing radiations are able to damage the molecules of this chemical, the chemical of genes, the cell may be made reproductively dead (not able to divide), or the cell may produce flawed "daughter" cells when its does divide.

If the radiations is having affected germ plasm, the flaws may eventually give rise to permanent genetic damage to offspring; if the damage has been in somatic tissue,they may give rise to tumors or cancer. Radiations in to bone marrow, for example can give rise to leukemia. Radiations into the pelvic region of a pregnant woman may cause damage to the fetus. X-rays and gamma rays penetrate intact skin ; alpha rays and beta rays cannot. All forms may bring about skin burns and years later skin cancer may develop. Massive absorbed doses induce a collection of symptoms called radiation sickness. Radiations enter the body on airborne, radioactive dusts and gases, or in food. The death rate from lung cancer in the infamous uranium mines of Joachimsthal, Bohemia was 30 times normal in the 1930s. Radon, a radioactive decay product of uranium and radium, has been a chemically inert but radioactive gas that did not dissipate readily in the mines, which have been worked for lead, cobalt, and arsenic, and the frequency of "lung diseases" among the miners was noted more than 400 years ago.

Table 9

Average LET Values and Tissue Penetration Distances for Electrons, Protons, and Alpha Particles.

1 keV = 1000eV (Electron Volts) = 1.6 x 10^{-16} Joules.

1 μm = $10^{-6}m$

Particle (keV)	Energy (keV/μm)	LET distances (μm)	Tissues penetration
Electron	1	12.3	0.01
	10	2.3	1
	100	0.42	180
	1000	0.25	5000
Proton	100	90	3
	2000	16	80
	5000	8	350
	10000	4	1400
	200,000	0.7	300,000
Alpha	100	260	1
	5000	95	35
	200000	5	20000

As radiation damage in a cell has been most serious when it affects the generic chemicals that are critical in cell division, the symptoms of such damage appear earliest among tissues whose cells divide most frequently. Tissues forming the walls of the stomach and intestines and the tissue that makes white blood corpuscles in bones have such cells. It has been not surprising, therefore, that early symptoms of radiation damage involve the stomach (nausea and vomiting) and intestines (diarrhea), as well as a drop in the supply of white blood cells.

Strontium 90 (half-life 27.7 years) has been dangerous because it has been a bone- seeker,being chemically similar to calcium, a component of bone. Cesium 137 (half life, 26, 6 years) is chemically similar to sodium, an element present as the sodium ion in ordinary table salt. Ions of cesium 137-are easily distributed throughout the entire body, including the gonads. From a biological point of view, radiation effects are also classified as somatic or genetic.

Genetic effects have been those imvolving mutations of the chromosomes or genes in sex cells; they pose a potential hazard to the descendants of the exposed person and are of concern to the whole society. Genetic effects have been of grave concern only in persons who have been not yet past their reproductive years.

We are lacking a basic understanding of the biological action of ionizing radiations despite a great deal progress in research. Most experience have beem carried out using laboratory animals, and comparison with humans has been difficult to make. Some data could be accumulated over the years from human exposures to large radiation doses obtained accidentally or from nuclear explosions.

Table 11 lists the approximate short-term effects that might be experienced for whole body adiation exposures over a short period of time. As different persons differ in their sensitivity to radiation, as in their sensitivity to infections and numerous other things, the doses corresponding to different effects will vary widely from person to person.

Exposure to a few hundred rads give rise to acute radiation illness: nausea, fatigue, and vomiting within a few hours and for a day or two, a decrease in red and white blood cells and blood plate-lets for a few weeks, then anemia, susceptibility to bacterial infection and hemorrhaging for some period of time, followed often by death.

Table 10
Estimated Short-term Effects of Single-dose, Whole-body Radiation Exposures in Man

Less than 25 rads	No observable effect
About 25 rads	Threshold level for detectable effect
About 50 rads	Slight temporary blood changes
About 100 rads	Nausea, fatigue, vomiting
200 to 250 rads	Fatality possible, though recovery is more likely
About 500 rads	Perhaps half the victimes would die
About 1000 rads	All the voctims would die.

If the victim is able to survive, he or she wiil still have a greater-than average chance of developing leukemia (especially in the first few years after exposure), other forms of cancer, cardiovascular disorders, and eye cataracts. Atom bomb survivors showed higher than average leukemia rates,which packed in 1951-6 years after exposure- but which were still higher than usual in 1966.

Mortality rates for persons within 1200 m of the hypocenter in the two towns, after excluding the effects of leukemia were 15% higher than in unexposed Japanese during the decade 1950 to 1960- a statistically significant increase. No effect had been found on mortality of children conceived to survivors after the exposure, however.

Very high rates of chromosomal abnormalities are found among survivors and their children who were in utero at the time of the bombings but not in children conceived afterward.

Despite frequent criticism of these studies in Japan, these studies are being continued and are playing an important role in defining the hazards of radiation. They are especially important now when increased cancer rates or other effects might be appearing after a long period of latency.

The injurious effects of radiations have been seen mainly in two forms. One is direct damage to the tissues ans organs of the exposed organism called somatic effects, from the Greek soma, meaning "body". The other is the indirect damage to subsequent generations through the disruptive action of radiation on the reproductive tissues of the parent, called genetic effects, from the Greel word genesis, meaning "descent".

Genetic effects are caused by the action of radiation on the molecular or physical structure of the chromosomes.Even in the production of somatic injury, the focal point of the ionizing radiation may be its disruptive action on the DNA make up of the chromosomes of the somatic cells. Rememnber that DNA is the vital substance that has the dual function of guiding the somatic cellular activities of the individual organism as well as prepetuating racial characteristics in the progeny.

Somatic Effects

The first evidence of the damaging effects of radiation came from unfortunate experiences of people who, because of their occupations, either used or came in contact with sources of high-intensity radiation and suffered chronic exposure. Radiologists, uranium mine workmen, and painters of radium dials were among those who suffered disastrous effects. Patients who received large doses of X-rays and those who given natural radionuclides such as radium and thorium for therapeutic purposes were also afflicted.

Additional evidence of the degree and kind of damage from radiation was forthcoming from studies of the Nagasaki and Hiroshima survivors, from the unforeseen exposure of the Marshall Island people following a weapons test in 1954, and from accidental exposures among laboratory and industrial workers. Finally, laboratory animal experimentation has been a rich source of information on the effects of ionizing radiation, dating as far back as 1927, when Hermann Muller demonstrated that x-rays can cause drastic changes in fruit flies. Many other investigators have since then shown damaging effects to animals and plants.

1. *High Radiation Exposure.* A lethal dose of penetrating ionizing radiation to humans (about 500 rads, see Table 1.12) causes an acute form of radiation sickness. Within a few hours the patient becomes nauseated and experiences vomiting and fatigue. In the typical radiation syndromes the symptoms last a day or two after which the patient feels fairly normal for a time. However, his blood is not up to par.

There is a gradual decrease in the number of red and white blood cells over a period of several weeks until it is obvious that anemia has developed. The blood has lost much of its capacity to carry oxygen and the patient feels weak. Because of the drop in white cells, he is more susceptible to infection. There is also a decrease in thenumber of blood platelets, essential for clotting. As a result, the patient begins to bleed in various parts of the body. There is hermorrhaging from the nose, gums, and intestimes.

Blood clots blotch the skin and mucous membranes. If there are wounds, they heal poorly. The victim declines in vitality until he dies from anemia, infection, and hemorrhage. Individuals that do not die within a few weeks are still not in the clear. There occurs a sharp decrease in life expectancy. Typically in mammals, such as mice, there is an interval of low mortality with a peak at about 3 months following exposure, after which the death rate increases until 9 months after exposure. Delayed deaths may occur for several months thereafter is the result of progressive deterioration from the acute effects of the exposure. During pregnancy, the fetus is more susceptible to radiation injury than adults, the effects depending on the number of days from conception.

Table 10
Estimated Doses for Varying Degrees of Injury to Man

Accumulated dose (R) or dose rate	*Period of time*	*Effects*
500 R/day	2 days	Mortality close to 100%
100/ Rday	Until death	Mean survival time approximately fifteen days
60/R/day	10 days disabilities.	Morbidify and mortality high with crippling
30R/day	10 days	Disability, moderate
10 R/day	365 days	Some deaths
3 R/day	Few Months	No drop in efficiency
0.5 R/day	Many months	No large scale drop in life span

Parts of the body differ in their sensitivity to radiation injury. The most sensitive tissues from acute dosages are the intestines, the lymph nodes, and the organs responsible for regenerating blood cells-the spleen and bone marrow. The intestinal wall partially breaks down and becomes inoperative in its normal function asthe main barrier to infection from the bacteria that inhabit the gut. The reduced number of leucocytes (white blood cells) that normally attack invading microorganisms are unable to cope with bacterial attack, and the body succumbs to overwhelming infection aggravated by the failure of the blood to clot because there are fewer platelets.

There is also partial asphyxiation from anemia caused by the disappearance of the red blood cells. A further effect of ionizing radiation is that it destroys the body's immune response. One of the body's defenses against extraneous substances, especially invading organisms, is a biochemical defense consisting of the rapid production of antibodies that neutralize the intruders. Thus, in addition to lowering resistance by the destruction of white blood cells, radiation impairs the ability of the body to combat infection with antibodies, leaving it vulnerable to further invading microorganisms.

2. *Fractional Dose.* The biological effect of radiation sometimes gets decreased when the total radiation dose gets divided into several exposure periods over a period of time. For example, the susceptibility of patients whose radiation had been fractionated over an 8-day period had been significantly less than those who received the same amount of radiation in a single dose. In a typical experiment using fractioned exposures, twice as much radiation had been needed to produce a characteristic response.

3. *Low-penetrating Radiation.* The effects of the low-penetrating types of particulate radiation have been typically less severe than from penetrating radiation. This was demonstrated by a mishap that followed a text explosion at Bikini Atoll in 1954.

Radioactive material was thrown up from a 15-

megaton thermonuclear device. Unpredicted winds at high altitudes swept the nuclear debris off its expected course and deposited radioactive material on Rongelap, an inhabited island about 100 miles east of Bikini. Radioactive debris also fell on a Japanese fishing vessel, the Lucky Dragon, that was operating in the area.

The natives of Rongelap got suffered exposure that required medical attention and prolonged observation. There were evacuated from the atoll and it was not safe enough for them to get returned until 1957. The Rongelap people received a heavy dose of beta radiation from radioactive material that settled on their skins. For a few days they got bothered with considerable itching and burning but there was no erythema (redness of the skin).

Then it about 3 weeks, patches of darkened skin and raised areas got appeared, especially on the scalp, and there occurred some loss of hair. Within a few months the injuries completely healed. With highly penetrating radiation like photons of x-rays, much less exposure would have been needed to cause serious injury.

4. *Delayed somatle Effects.* If the animal or human patient survives the radiation sickness from acute exposure, the possibility of illness or death remains from one or more of several profound effects on the cells and tissues. The foremost delayed somatic effect is the increased chance that cells will become cancerous, but he appearance of cancer may be delayed for months or years.

The type of cancer most commonly associated with radiation, and the first to appear, is leukemia, a cancer of the blood-forming tissues in which the white cells increase in number and become maligant. Leukemia may show up about a year after irradiatis, or even later. The chance of eventually developing one of the many other forms of radiation-induced cancers is estimated to be about five times that of leukemia.

There is also the increased probability that the victim will be afficted later in life with anemia, various cardiovascular disorders, or eye cataracts, as well as premature aging and reduced longevity. If the organism is young, he will be stunting or other impairment in growth. Fertility is reduced or temporasily impaired. Exposure of the female ovaries to a dose of 300 rads or the male testes to 1,000 rads brings about permanent sterility. Delayed effects showed up among the residents of Rongelap Island, which received fallout from the Bikini test of 1954. The sixty-four residents of Rongelap got exposed to both gamma and beta radiation. The dose was not lethal but caused nausea and vomiting, and there was a reduction in blood cell elements that lasted for several months. Besides the skin exposure that resulted in "beta burns" and loss of hair, there was an intake of radioactive iodine with contaminated food and water.

The iodine was deposited in the thyroid glands and this proved to be the most serious aspect of the exposure. Healing of the skin and regrowth of hair, as well as nearly normal recovery of the formed blood elements, were complete by the end of the first year. The effects on the thyroid glands were longer lasting. Observers noted abnormalities of the thyroid glands ten to fifteen years after exposure.

It has been highly probable that the effects had been the result of deposition of radioactive iodine in the thyroid at the time of fallout. Ninety per cent of the thyroid abnormalities had been in children who were under ten years of age at the time of exposure. Some of the children were retarded in growth and development, an effect that got related to the thyroid injury,

5. *Life Span and Aging.* Experimental work on animals reveals that radiation shortens the life expectancy of rats, mice, dogs and other animals. The causes of reduced life span are not confined to cancer but extend generally over the normal spectrum of diseases. From such data, it was estimated that at low levels of radiation exposure, the life expectancy of humans may be reduced by as much as 5 days per roentgen. This implies that, during a normal human life span, an average exposure to man-made radiation of six rem (or roentgen) would cut the life span by about 1 month.

The effect is usually proportional to exposure, so the life-shortening effects of higher exposure rates, such as from excessive diagnostic or therapeutic x-rays, would be proportionately greater. Other delayed somatic effects opf radiation have been cataracts, sterility or impairment of reproduction depending on exposure rate, and accelerated aging.

These have all been demonstrated in laboratory animals. The cancer-causing effects of radiation have been extensively documented for both laboratory animals and humans. Some radiation workers believe that diagnostic radiation of pregnant women may be able to increase the risk of cancer to the child, although experiments with fetal mice show them to be surprisingly resistant to radiation before birth.

Genetic Effects

When Hermann Muller x-rayed his fruit flies and found that the mutation rate was thereby greatly increased over that which occurred from the effects of the natural environment, his discovery made it possible for geneticists to study large numbers of mutations in the laboratory. But beyond that, it removed much of the could of maystery that had surrounded mutations. It showed. Muller reasoned, that mutations had been the result of changes within the genes brought about by conditions in the environment to which the flies were subjected.

Furthermore, these changes could be initiated by human intervention. Muller set the stage for the subsequent discovery by Charlotte Auerbach that not just radiation but ordinary chemicals as well could cause gene mutations. Twenty years after his break-through discovery in 1926, Muller was finally awarded the Nobel Prize in medicines and physiology. Later, other geneticists found that mutations could be brought about by different forms of radiation.

1. *Mutation Rates.* Mutations in the human population have been difficult to study due to the long human life span, the comparatively small number of offspring, and fact that the human mating and reproductive habits are almost impossible to study under anything approaching controlled conditions. Geneticists have turned to small organisms that are able to reproduce rapidly and can be grown in the laboratory under carefully controlled conditions. The common fruit fly, Drosophila melanogaster, is having only four pairs of chromosomes. The chromosomes could be identified and mapped easily and the locations of the genes can be determined. This insect has served science admirably in divulging much that we know about mutations, as well as inheritance in general, from the time the American geneticist Thomas Hunt Morgan started his experiments with the insect in 1907.

Morgan and his students found that it was possible to pinpoint on a chromosome the spot at which a gene for a particular characteristic was located. Soon they were drawing up "chromosome maps" for their fruit flies. It was on this foundation that Hermann Muller, a student of Morgan's, was able to speed up the rate of natural mutation in fruit flies in ways (especially with x-rays) that enabled him to observe the deleterious effects of mutations. There has been little that we can do at present to protect ourselves from natural radiation, but it has been important to know what the effect is going to be when expose ourselves to additional ionizing radiation.

Mutations are generally harmful because a random change in the molecules of the genes is very unlikely to result in a more satisfactory information being coded on the genes. However, once in a while a desirable change may be produced in a gene. Then it is through the processes of successive selection and adaptation during which the off-springs produced due to deterious mutations do not survive, that the off-springs with beneficial mutations finally stabilise after several generations.

Natural mutations have been taking place since the origin of organic life on the earth. They are known as spontaneous mutations. They may be produced by several methods and exposure to nafural radiations is one of them. Muller showed in 1927 that the rate of mutations could be controlled by exposing fruit fly (Drosophila M) to X-rays. The off-springs were defective as they lacked wings. Similar experiments with other types of ionizing radiations on animals and plants have confirmed these results which are similar to those due to spontaneous mutations and mutations caused by mustard gas and other chemicals.

The increase in frequency of mutations is found to be proportional to the total dose received by the gonads of the species. There are indications that some repair mechanism also functions at low doses to reproductive organs before the mutations actually occur in the germ cells. But once these cells have been damaged there is no possibility of a recovery.

Increased mutations due to radiation exposure can produce individuals with increased susceptibility to disease, shortening of life, sterility, clour blindness, mongolism, and other defects. Individuals with more harmful genes are expected to be eliminated rapidly while those with less deleterious genes may survive to reproduce a large number of defective off-springs.

Experiments conducted by Luning by exposing the sexual organs of mice with a dose of 2 Sv for nineteen generations showed no genetic defects. Again an investigation of thirty six thousand children born to parents, who had received nearly 3 Sv dose during stomic bomb attacks on Japan, have shown no increase in genetic defects over those of children born to parents who did not receive radiation exposure.

Results on friut fly have been used to estimate that a dose of 0.3 to 1 Sv to gonads will double the natural mulitation rate. However, it is very unlikely that the results on fruit fly are directly applicable to man. A more realistic estimate is that the genetic defect frequency per live birth per 10 mSv of dose received by father is 1.5×10^{-4}. Accordingly in a population of ten million with a birth rate of 2 per cent the number of persons both with genetic defects when all fathers are given an excess dose of 10 mSv will be 30.

2. *Effect of background radiation.* It was said earlier that various types of radiation, generally of low intensity, are part of the natural environment. This is called background radiation. The earth and all its plant and animal inhabitants are bombarded daily by high-energy particles from the sun and outer space. Rocks and soil contain radioactive isotopes of many of the elements in variable but usually small amounts. Radioactive elements are picked up by water and are taken into plants along

with mineral nutrients, thereby gaining entry into the food chain.

Food itself varies in radioactive content, and even the tissues of our bodies inevitably contain sources of radiation in varying degrees depending upon the concentratin of radiation sources in the food, water and air. The amount of background radiation is not the same everywhere on the earth. Radiation from space is partly absorbed by the atmosphere so that less of it reaches sea level than mountain tops. People who live in the mile-high environs of the Rocky Mountain Plateau may be exposed to twice as much radiation as those who inhabit the Atlantic or Pacific coasts.

There is also more intense radiation in the polar regions than at the equational belt because cosmic rays streaming in from space tend to be pulled toward the poles by the earth's magnetic field. The earth's magnetic field stops some of the radiation resulting in high-intensity radiation in the so-called Van Allen zone. The amount of radiation from minerals in the soil varies greatly from place to place. For example, the inhabitants of Kerala, India are exposed to many times the average amount of environmental radiation because of the high radioactive thorium content in nearby soil.

Table 11

Sources of Radiation Exposure

Source	*Average dose per year, millirems*	
	Whole body exposure	*Genetically significant exposure*
Natural radiation		
Cosmic radiation	44	
Radionuclides n the body	18	

External gamma radiation	40	
Subtotal	102	90
Man made radiation		
Medical and dental	73	30-60
Fallout	4	
Occupational exposure	8.0	
Nuclear power (1970)	0.003	
Nuclear power (2000)	< 1	
Miscellaneous	2	
Subtatal	80	
Total	182	
Radiation Protection Guide for man-made radiation (medical excluded) to the general population	170	

For this purpose the rem, roentgen equivalent mammal (or man), is used as an expression of the relative danger from different sources of radiation. For example, it is estimated that for the average person the gonads (reproductive organs) of the body are exposed to about 3 to 4 rems during 30 years. This span is taken as the time for ine generation during which the human reproductive organs will undergo development plus the most active reproductive years.

The National Academy of Science estimated that in the United States the average genetically significant exposure of the gonads to radiation from natural sources is 90 millirems (thousands of a rem) per year. Of this, about 80 per cent is from sources outside the body, the remainder from elements found naturally in human tissue and from inhalation of air.

3. *Man-made radiation.* Ever since the discovery of X-rays and the isolation of radioactive chemicals in the

laboratory late in the last century, people have been increasingly exposed to ionizing radiation from its widespread use in medicine and industry.

The nuclear age, beginning with the development of the atomic bomb in 1945, propelled the environmental burden of radiation to new heights, not only from the explosive release of radionuclide in the atmosphere which eventually invade water, soil, plants, animals, food, and the human body, but also from the greatly increased use of radioactive substances in medicine, industry, and research.

Life-time of records of exposure to man-made radiation are almost non-existent, although the practice of keeping such records has been strongly recommended. The Federal Radiation Council estimated that of the average exposure of people in the United States to all man-made sources (67 millirems per year), 75 per cent is from diagnostic x-rays. Of the remainder, about 16 per cent is from radiation therapy, 6 per cent from radioactive fallout, 3 per cent from such miscellaneous sources as luminous watch dials, television tubes, and industrial wastes, and a small additional amount from nuclear research laboratories.

Thus most of the man-made radiation- more than 90 per cent- is from medical uses, and this comprises about 30 per cent of all radiation to which people are exposed, both natural and man-made. If the higher value is taken as the exposure that might be expected in the United States, equal to 5.5 rem (5.500 millirems) for a 30-year period, the effect of man-made radiation is now nearly twice that already caused by natural, or background, radiation.

4. *Radiation Effects on the Gonads*. An exposure of reproductive organs (gonads) to ionizing radiations can

result in damage to the germ cells which take part in the reproduction process. In normal cells the zygote formed from the fusion of sperm and egg cells from two parents in a normal one. Its growth and subsequent division to duplicate gives cells in which the harmosomes and the genes are copies of those derived from the two parents.

As a result the off-springs have features which are derived from the genes pf parents. However, if the germ cells have undergone molecular changes in the genetic material a mutation is said to have occurred. If such a germ cell takes part in a reproduction processes the characteristics of the off-springs will be different in some respects from those of the parents. These changes are called genetic mutations.

Genes are generally classified as dominant and recessive. In case the affected gene is a dominant one then the off-spring will contain the characteristics controlled by that gene. A recessive gene on the other hand must be present in both the parents to produce any effect in the off-springs. Normally a recessive gene will be dormant till the time both the egg and sperm cells have that particular gene.

General Biological Effects of Ionizing Radiations

The greatest danger of ionizing radiation has been that it has been insidious. Specialised instruments are necessary to reveal its presence. Damaging exposure can thus take place without awareness of it at the time. As the biological damage, which may be irreversible, is done at the instant of exposure, it is too late to do much about it by the time any symptoms appear. The dangers from ionizing radiations have been of quite different nature from radiations encountered more commonly such as uv, visible radiation etc. The general biological effects of radiations will depend upon the exposure and may include the following:

Effects on Individual Cells

Retardation of cell division, for a relatively small dose of radiation (50 to 1000 rads), delay in cell division has been an observed fact. The cells usually recover from this damage, if the dose of exposure have been not high; fraction of cell recovery depends upon the exposure dose. It has been established that the cells have been not equally susceptible to radiation during all of their growth phases; the sensitivity is found to vary depending not only upon the nature of radiations but also on the nature of the cells under study. The phenomenon of the delay in progression of mitosis is known as 'mitotic delay' produced by radiation.

Cell death. Ionizing radiations cause a special type of cell killing which is characterized by the loss of reproductive integrity, that is, the capacity of sustained cell division. The radiations have been, however, also capable of inducing death in the sense that we understand, but this has been produced only at a dose which has been larger by a factor of 10μ compared to the dose needed for reproductive death. It has been easy to explain death when very high doses are used; because, then degradation cellular components on a wide scale has occurred.

When the reproductive death take place, other cellular functions except the division ability, on the whole, remain unimpaired. It is reported that generally the number of cells killed by a definite dose of radiation exposure has been proportional to the number of cells present.

Chromosome breaks. The chromosomes have been dynamic entities within the cell and their structure has been not permanent. Irradiation makes the breakage of chromosomes, which may occur at a fairly low dose of irradiation, depending upon the stage at which cells are exposed. Chromosome breakage may cause serious consequences for the cells because chromosomes carry the genes. So, not only the molecular integrity but also the order of genes in the chromosome is important.

Effects on Human Being

Radiation effects have been of two types : (a) a large dose given in a short time, creating acute radiation sickness and (b) a small dose given over a large time causing the chromic effects. In case of chromic irradiation, biological effects may get delayed for periods from a few days to many years depending upon the dose and type of radiation received. Some of the common physiological effects have been described as follows:

Leucopenta. Continued over-exposure to penetrating radiation first get exhibited by a decrease in the total white blood cell count, called leucopenia. The decrease has been mainly due to a reduction in the absolute number of polymorponuclear cells, whose recovery power has been not as great as that of the lymphocytes. Therefore, the lymphocytes will represent a larger percentage of the total number of white cells present, instead of the normal 20 to 30 per cent. This condition, called relative lympocytosis has been also observed in a transient slale about 8 days after an acute whole body irradiation of adequate dose.

A low red cell count(anemia) due to chronic over-exposure is seldom seen, probably due to the greater radiation resistance of the red cells and of the erythropoietic system in which are formed.

Epilation. X-ray exposures can give rise to epilation and there was a time when x-ray exposures were used to remove superfluous hair, not without injury from the large dose required. In some cases hair may return after an interval of some months but with a large exposure dose, epilation has been permanent. One of the prominent aftermaths of atomic bomb explosions had been the excessive epilation among people exposed to penetrating radiation.

Sterility. X-rays can cause temporary or permanent sterility. Temporary sterility could be produced with moderate doses and normal functions may return in a few months. Extremely large doses of penetrating radiation may cause permanent sterility.

Tumour production. Radiation is widely used in the treatment of cancer, but cancer may follow because of excessive radiation exposure.

Radiation Effects on Biological Molecules

Macromolecules may undergo structural alterations, which could be manifested as a change in physical properties affording sensitive means of detecting a reaction affecting only one or two atoms in a molecule containing several thousand of atoms though the chemical nature of the change cannot generally be determined at this level.

The formation of chemical bonds has been another effect of radiation. In dilute solutions, it has been predominantly intramolecular, and may result in the molecule being pulled together to occupy less space, so that the viscosity of the solution gets produced. At higher concentrations of the solute, cross-linking has been mostly intermolecular between different molecules.

As the radiation dose increases, the number of cross-links increases, and the solution gradually becomes more viscous eventually turning into a gel. Breakage or degradation of a macromolecular may effect the primary structure, resulting in pieces of various sizes. Such main chain scission often takes place at one particular bond, suggesting that energy gets transferred along the chain from the site where it was absorbed to the weakest link.

Disruption of the secondary and higher orders of structure may also occur, e.g., through the breakage of disulphate bridges, as in proteins, or very often by the rupture of weak hydrogen bonds.

Proteins. With proteins; direct action of radiation has been considerably more efficient than indirect: as little as 50-200 eV of energy deposited directly in a protein can inactivate it while some 50-200 hydroxyl free radials are needed indirectly, although one radical may be enough if it hits the right place. Indirect action may also produce important changes, e.g., insolubility. All enzymes could be inactivated in solution, the doses required vary widely, and some protection is afforded by sulphur compounds.

However, ateration of protein structure and function does not seem to be greatly important in primary radiation effects on biological systems. *Nucleic acids.* In nucleic acids, several significant changes occur, both in vitro and in vivo through in the cell there has been some protective effect from the proteins associated with the nucleic acids. Breakage of hydrogen bonds has been likely to be significant only if several have been broken at once in the same region the inherent rigidity of the structure, unlike that of proteins, holds it together while hydrogen bonds are brok en and allows them to reform correctly.

Breakage of the main strands may take place; a break in one chain only will be difficult to detect as the DNA helix is inherently rigid and will hold together. Breaks may join up to form cross-links, within the same DNA molecule, between two DNA, or between DNA and protein. If there occurs a breakage in the chain in less than about five nucleotide units of a break in the other, a 'double break' arises and the molecule separates into two pieces because there are now too few hydrogen bonds between the breaks to hold it together.

Such a double break arises much less commonly than a breakage in one chain only; it has been estimated that there has been a 1 in 70 chance that two single breaks formed at random will come close enough together to make a double break. Alternatively, a single particle with high LET may cut through both chains at once, for which about 600 eV get required.

Lipids. The chief reaction of lipids concerns the polyunsaturated fatty acids, and affects the carbon atom between the double bonds: hydrogen would be removed and a sesonating structure gets formed Peroxy1 radicals and organic peroxides are radily produced in the presence of oxygen, often by means of free-radical chain reactions affecting from 20 to 1000 molecules.

Carbohydrates. Radiochemical studies of carbohydrates have been few and inconclusive. Irradiation of sugars in dilute solutions, in the absence of oxygen, causes polyment production. This may possibly be due to cross-linking. Extremely high doses can produce degradation also.

A clue to the nature of the initial radiation lesion is given by the fact that a dose of 100 rad kills many cells, but inactivates only a few molecules in those cells, most being completely unaltered. It is also known that the RBE of radiation in vivoincreases with increasing LET though the RBE is not the same for different effects. The properties of the lesion must agree with these observed facts, and a number of possible alternative hypotheses have been proposed.

DNA complexed with proteins has been considerably less sensitive to base destruction or chain breakage by radiation than has been non-complexed DNA. Part of this 'protection' is from competition for radicals, but a large fraction is also due to the structure of the DNA protein complex. Protein gets wrapped around the DNA giving a shielding effect.

Key metabolic enzymes are less likely to be involved, since there are relatively many molecules of them, and they can be shown to be still functional after a damaging radiation dose. Interference with DNA is almost certainly the most important cause of long-term and genetic radiation effects. Radiation effects might be able to follow the interruption of energy supplies: blocking of key enzymes might stop normal metabolism, so that the body would have to use other energy sources which might be quickly exhausted.

Radiation might be able to position the cell, either by direct alteration of particular compounds, or indirectly via the interruption of energy supplies or by the accumulation of intermediary metabolites that are toxic of high concentration. Kuzin et. al. postulated that the oxidation of orthophenols to orthoquinones has been fundamental in the biological action of high energy radiation. Glyoxal may be a cytotoxic agent resulting from the irradiation of carbohydrates.

Radiation might break down intracellular barriers. This the so called membrane damage or enzyme release hypothesis, postulated that the primary lesion caused by radiation had been the damage to the membranes around and within the cell, which could resist isolated ionizations but not the impact of several ionizations simultaneously. The consequent alterations in the structure of the membranes lead to changes in their permeability or to their rupture when sufficiently extensive. Thus all kinds of molecules which are normally held on one side of the membrane can escape, including enzymes, which are then able to react in uncontrolled manner with subtrates they might not normally encounter. This theory was able to explain better than others many of the observed immediate effects of radiation, and in this respect is now favoured as being the most likely to be true.

Through it, the former nucleus versus cytoplasm controversy over the location of the primary site of radiation damage has in part lost its meaning for, once intracellular barriers are broken, the released enzymes are free to travel, and the site of the primary lesion can no longer be identified with the site at which the enzyme causes damage. Cytoplasmic enzyme can act on the nucleus, because the nuclear membrane allows the passage of large molecules.

Radiation might induce the release of virus. Fischinger and O'Connor (1969) provided evidence that x-irradiation can induce a leukaemogenic virus in mind. But no one hypothesis could satisfactorily explain for all the observed effects of radiation on the cell, and it is likely that all the mechanisms of damage in the several theories have some part to play at some point in space and time.

Effects of Radiations at Molecular Levels

One of the principal causes of cellular death has been the DNA damage which is induced by ionizing rays and ultraviolet light. The study of radiation induced damages in DNA thus becomes important. The effects could be appreciated in terms of the DNA structure of the Waston and Crick.

The most important feature of the structure of DNA molecule has been that it is having two polynucleotide chains which are wound up together to form a double helix and held together by specific bonds between purines and pyrimidines in different chains (Fig. 16) The general features of DNA structure depend mainly on two main types of chemical bonding viz., the covalent polyester bonds between the nucleotides which form a single polynucleotide chain and the hydrogen bonds between the purines and pyrimidine ring systems of two different polynucleotide chain which produce and stabilize the double helical structure.

The following chemical effects have been observed to occur when DNA gets irradiated with ionizing radiations in dilute solutions: breakage of phosphodiester bonds, oxidation of sugar moiety, deamination and dehydroxylation, formation of hydroperoxides. 4-5 double bonds in pyrimidines have been sites susceptible to attack by free radicals. Purines have been more stable to radiation degradation by a factor of two. Of the four bases, thymine is most susceptible to degradation. Experimental values indicate that a phosphodiester backbone scission is just as probable an event as a base destruction and that both reactions involve simultaneous destruction of sugar ring. The macromolecular integrity of nucleic acids get destroyed on breakage of the sugar phosphate backbone. The average molecular weight could get reduced on introduction of one scission per strand in any single stranded RNA or DNA molecule.

Thus at low radiation dose at the level of one or two hits on the linear molecules viscosity and sedimentation rate of the polymer get altered. In circular single stranded DNA one backbone scission converts the circle too an open liner structure without change in molecular weight. A clear separation of the open and closed hain polymer can be achieved by hand centrifugation, and the kinetics can be followed easily as function of radiation dose. By constrast, single-strand scission in a linear double-stranded DNA, radiation-induced or otherwise remains hidden due to the co-operative hydrogen bonding between the two polynucleotide strands.

It is reported that as many as 100 DNA seinduced single strand scissions can remain hidden in a DNA molecule of the size of 10^7 daltons (Ginoza, 1967). A decrease in molecular weight has been detectable only on a further accumulation of these hits. This reveals that hydrogen bonds between non-coincident breaks in opposite chains can resist thermal energies at the room temperature (25°C) and maintain a linear integrity of the polymer.

The degradation induced by x-ray and r-rays and has been found to be proportional to the square of the radiation dose, indicating support for two independent-hit mechanism. Any past-irradiation manipulation that causes hydrodynamic shear has been likely to break the DNA molecule at a site oppsite a single strand break which is a point of weakness. The region between two close non-coincident breaks inopposite chains of DNA, also a region of weakness, being more susceptible to hydrogen bond breakage. It is now possible to detect even an average of one single strand break per 10^7 molecular weight DNA molecule with the refined techniques of ultracentrifugation.

A classical work done nearly three decades ago provided evidence that ionizing radiations inhibited DNA synthesis.Using heat denatured. DNA as a primer in the calf-thymus polymerase system. Setlow and his co-workers (1968) reported that thymine dimers induced in DNA by ultraviolet light irradiations block template function. Because the lesion induced by ionizing radiations are heterogenous, no given lesion can be clearly identified as the cause for similar loss in function is known to occur when DNA has been x-irradiated. In any case, one expects that DNA, the primary template from which all cellular attributes stem, must be held inviolate so that corect-information be passed into the cell and its progeny.

In vitro experiments reveal that the dose of ionizing radiation required to destroy the template function of DNA (mammalian), for DNA synthesis is at least an order of magnitude higher than for mRNA synthesis. The D_{37} for the two functions measured are 1×10^5 rad and 5x10 rad, respectively.

The function of DNA is isolated calf thymus nuclei for DNA synthesis has been reported to be much more radio-resistant than the template DNA irradiated in solution. In vivo testing of the radiation sensitivity of the template functions of DNA gets complicated by the fact that when a bacterial cell is exposed to radiations the DNA has been known to undergo extensive degradation, depending on the genetic trait of the cell involved.

The dimers formed between adjacent pyramidines by ultraviolet light have been chemically the stable and well-defined lesions readily produced in DNA.

The breaking of polynucleotide chains may be having devastating biological results, but such breaks appear only at higher doses and has been oxygen dependent. But lethal action of uv has been not oxygen dependent. Therefore, the chain breakage may not be a major factor.

If thymine dimers get involved in biological damages, the activity lost by irradiation with 280 nm should at least get partially reversed by 239 nm. Such effect has been observed in DNA priming activity in a calf thymus polymerase system. The activity has previously been destroyed by 280 nm. This reveals but does not prove that thymine TT dimers are involved, since along with thymine dimers other photochemical products are also formed. If we are able to separate the photochemical changes into dimers and other unknown products X and let T∧T equal the fraction of TT sequences which are dimers, we may be able to represent the survival S, as function

$$S = F[(X); (TT)],$$

In the large dose range the survival curves have been very roughly exponential and the contribution of (X) and T∧T to inactivation may be found from the relative slopes of the survival curves. If K_1 is the relative slope for 280 nm inactivation alone and $K_{2'}$, for 280 nm+239 nm inactivation, the relative contribution of T∧T to inactivation is $K_1 - K_2/K_1$. Such analysis indicates that 50% of the inactivation at high doses arises from the formation of thymine dimers.

The irradiation of DNA brings about absorbance changes in the spectral range 220 to 330 nm. The changes have largest at 270 nm. Moreover the individual nucleotides have about equal absorption coefficient at 270 nm, thus making subsequent analysis simple.

Alkylated bases. The nature of the principal products formed by certain alkylating agents such as methyle methane sulfonate (MMS) ethyl sulfonate (EE), nitrogen mustard, $-NH_2$, and sulfur mustard has been determined. It was proved that the N7 position of guanine and N3 and N7 positions of adenine have been the most reactive sites toward alkylating agents in aqueous solutions of free bases. For example, the principal products formed by treatment with MMS include 7-methyl-guanine and 3- or 7-methyladenine. In native DNA, however the most frequent reaction is the alkylation of N7 atom of guanine.

Di-functional alkylating agents like-NH_2 of sulfur mustard produce both guanyl and di-guanyl products. These are generally formed by reaction with the N7 atom of guanine, which is sometimes followed by a slower reaction which crosslinks to a second guanine. There exists a tendency for the alkylated purine moieties to hydrolyze slowly, leaving an apurinic site, a process which may take many hours depending upon the particular agent used.

It was demonstrated that many of the diguanyl products fromed in native DNA by treatment with sulfur mustard consitute cross-links, between strands. It was reported that about 13 alkylations by sulfur mustard in the DNA of bacteriophage T4 were required to produce a lethal event.

In contrast, over 200 alkylations were needed in comparable experiments with a monofunctional alkylating agent. Thus, di-functional alkylating agents have been generally more effective than monofunctional in killing viruses or cells. Several other agents have been known to cross-link native DNA. Mitomycin C and profiromycin cross -link complementary DNA strands. After heat denaturation and quick cooling the strands renature without delay. Nitrous acid cross-links DNA.

In photodynamic action, dyes, such as methylene blue have been reported to sensitize cells to blue light in the presence of oxygen. This action has been attributed principally to the destruction of guanine.

Criticality

Ponds often have enough fissile material to 'go critical', that is to initiate a fission chain reaction. The hazard is controlled by. storage in safe configurations, that is with adequate spacing between fuel elements or by limiting the number of elements which may be out of their storage position at any time.

Loss of Shielding

A typical storage pond may often have several magacuries of fission product activity. Such an element unshielded would give a gamma dose rate in excess of 100 rem/h at a distance of 10 feet. Obviously this situation must not be allowed to take place.

Loss of shielding can take place by loss of pond water or by raising fuel too close to the surface of the water. Loss of pond water has been more likely to be due to accidental pumping out than to a leak caused by serious structural damage. The possibility of pumping out could be reduced by good design of the water system any by administrative control of the system (e.g., locks on vital valves).

Operational Aspects

The fission product activity of the fuel is present within the fuel cladding. A breach of the cladding would give rise to contamination of the water by fission products to some extent depending on the form of the fuel.

As a precaution against the release of gaseous fission products, many ponds are having an air extract system 'sweeping' the water surface. Some degree of water contamination also takes place due to the release of radioactive corrosion products from the surface of the fuel cladding. In order to prevent build up of activity a water treatment system is provided. Cutting operations in a pond can give rise to serious contamination problems due to swarf, particularly if fuel is cut.

Pond Instrumentation

There are three main types of instrument which are used to give warning of hazardous coditions:

(i) Installed gamma monitors constitute an essential feature of any fuel storage pond. A local alarm like a bell or buzzer should sound if any of the instruments reach some quite low level, say 10 mR/h. It may warn the operators that a potentially dangerous situation has been developing. At some quite high radiation level, say 10 R/h at the operators position, an evacuation alarm should sound. This alarm should work on a two out of three or two out of four basis to avoid spurious evacuations.

(ii) Pond Water Counter is a simple device. It continuously monitors the water for beta activity and provides early warning of damaged fuel. A shielded liquid flow Geiger-Muller tube connected to a ratemeter has been generally used.

(iii) Air monitor is a continuously operating airborne particle monitor in the general area or in the air extract. It gives warning of high airborne contamination which arises from pond operations or maintenance of pond equipment.

Transport of Irradiated Fuel

The transport of irradiated fuel is subject to the provisions of the Nuclear Installations Act. The main considerations have been the prevention of:

(a) Criticality.

(b) Loss of shielding, and

(c) Damage to fuel resulting in a release of fission products.

These hazards get minimized by the use of transport containers of a very high standard, capable of withstanding serious accidents.

Radioactive Waste

On a much smaller scale, radioactive wastes arise in hospitals, factories, research and teaching institutions because of the many applications of radiation. In such cases, the complex treatment plant used at nuclear power stations would be much expensive and so disposal may be done by the normal refuse collection or sewage systems.

High-level nuclear waste consists of the intensely radioactive used fuel elements and other material from nuclear reactor cores. Its radioactivity arises in part from fission products (e.g., ^{96}Sr and ^{231}I) which are short-lived but have special biological dangers, and in part from actinides, heavy metal of often very protracted radioactivity.

Since used fuel elements contain a good deal of radioactive fissile material, including unconsumed fuel, reprocessing is likely to be an important part of any long-term nuclear waste disposal strategy. Lower-level nuclear wastes will include all the lightly active abjects from reactor environments—tools, soiled protective garments, etc., and even perhaps run-off rainwater—all of which need to be taken into account in assessing reactor safety.

The importance of low-level waste is controversial: it can affect the population as a whole not just workers in the nuclear industry, but it seems at present to be negligible compared with natural sources of radioactivity. On the other hand it will have to be controlled with increasing care if nuclear-power use expands in the future.

There are three basic principles which may be applied in dealing with radioactive waste:

(i) Delay and decay,

(ii) Concentrate and store, and

(iii) Dilute and disperse.

The method or methods employed in a given situation is dependent on many for factors, example the quantity and type of radioactivity, its physical and chemical form, the geographical location. The more useful classification of waste material has been in terms of its physical form, that is solid, liquid or gaseous waste.

Consequences of Releases of Radioactivity

Any release of radioactivity has been a potential source of radiation exposure to the population at large. The radiation exposure can occur via many different exposure pathways. A release of activity from a chimney stack would be dispersed by air movements and could result in radiation exposure of the population in a number of ways:

(i) direct external β or γ radiation from the plume,
(ii) inhalation of radioactivity resulting in internal dose,
(iii) direct exteranl β or γ radiation from deposition (fallout) of radioctivity,
(iv) consumption of foodstuffs (e.g., vegetables) contaminated by deposition, and
(v) consumption of meat or miik from animals which have grazed on contaminated ground.

Similarly, radioactivity discharged into a river, lake or sea could give rise to a number of exposure pathways:

(i) contamination of drinking water supplies,
(ii) external dose to swimmers or to people using contaminated beaches, and
(iii) consumption of contaminated fish, shelfish or plants.

Some of these exposure pathways result from complex routes called food chains. For example, activity discharged to the sea may be taken up by plankton which are easten by insects, which are in turn eaten by fish. The fish are subsequently eaten by man.

Storage of Radioactive Waste

Storage of radioactive waste has been regarded as a very

useful procedure when dealing with nuclides of relatively short half—life (e.g, up to a few months). If we keep the waste for a period of up to a few years facilitates disposal either because the activity would get decayed completely of is now of sufficiently low level to permit release.

The principle is most commonly used for liquid effuents in which a number of tanks are available and when one is full it has to be isolated until the activity gets decayed to an acceptable level for release.

When long-lived activity is involved, storage can only be considered as a temporary solution. At present, the practice has been to concentrate the solutions by evaporation and then transfer them to special tanks in large concerts vaults.

The half-life of many of the nuclides would be such that it will be many centuries before it will be safe for a person to enter the vaults. The problem is really being passed on to future generations. An important factor in the storage of high level waste has been the cost. Extremes and expensive precautions have to be taken to prevent any release of the stored radioactivity. Many new methods of disposal of fission product wastes are being investigated in many countries.

One of the most promising methods has been solidification of the waste and disposal in deep underground caverns or disused mines. Salt mines are considered to be particularly suitable as salt deposits do not contain water which transfer the activity back into man's environment.

3 Convention on Early Notification of a Nuclear Accident (1986)

The Convention on Early Notification of a Nuclear Accident was adopted by the General Conference at its special session, 24-26 September 1986, and was opened for signature at Vienna on 26 September 1986 and at New York on 6 October 1986. It entered into force on 27 October 1986, i.e. thirty days after the date (26 September 1986) on which three States expressed their consent to be bound by the Convention, as required under Article 12 thereof.

The text of the Convention, taken from a certified copy, is reproduced herein for the information of all Members.

The States Parties to this Convention,

Aware that nuclear activities are being carried out in a number of States,

Noting that comprehensive measures have been and are being taken to ensure a high level of safety in nuclear activities, aimed at preventing nuclear accidents and minimizing the consequences of any such accident, should it occur,

Desiring to strengthen further international cooperation in the safe development and use of nuclear energy,

Convinced of the need for States to provide relevant information about nuclear accidents as early as possible in order that transboundary radiological consequences can be minimized.

Noting the usefulness of bilateral and multilateral arrangements on information exchange in this area.

Have Agreed as follows:

Article 1

Scope of Application

1. This Convention shall apply in the event of any accident involving facilities or activities of a State Party or of persons or legal entities under its jurisdiction or control, referred to in paragraph 2 below, from which a release of radioactive material occurs or is likely to occur and which has resulted or may result in an international transboundary release that could be of radiological safety significance for another State.
2. The facilities and activities referred to in paragraph 1are the following:
   (a) any nuclear reactor wherever located;
   (b) any nuclear fuel cycle facility;
   (c) any radioactive waste management facility;
   (d) the transport and storage of nuclear fuels or radioactive wastes;
   (e) the manufacture, use, storage, disposal and transport of radioisotopes for agricultural, industrial, medical and related scientific and research purposes; and
   (f) the use of radioisotopes for power generation in space objects.

Article 2

Notification and information

In the event of an accident specified in article 1

(hereinafter referred to as a "nuclear accident"). The State Party referred to in that article shall:

(a) forthwith notify, directly or through the International Atomic Energy Agency (hereinafter referred to as the "Agency"), those States which are or may be physically affected as specified in article 1 and the Agency of the nuclear accident, its nature, the time of its occurrence and its exact location where appropriate; and

(b) promptly provide the States referred to in sub-paragraph (a), directly or through the Agency, and the Agency with such available information relevant to minimizing the radiological consequences in those States, as specified in article 5.

Article 3

Other Nuclear Accidents

With a view to minimizing the radiological consequences, States Parties may notify in the event of nuclear accidents other than those specified in article 1.

Article 4

Functions of the Agency

The Agency shall:

(a) forthwith inform States Parties, Member States, other States which are or may be physically affected as specified in article 1 and relevant international intergovernmental organizations (hereinafter referred to as "international organizations") of a notification received pursuant to sub-paragraph (a) of article 2; and

(b) promptly provide any State relevant international organization, upon request, with the information received pursuant to sub-paragraph (b) of article 2.

Article 5

Information to he provided

1. The information to be provided pursuant to sub-paragraph (b) of article 2 shall comprise the following data as then available to the notifying State Party:
   (a) the time, exact location where appropriate, and the nature of the nuclear accident;
   (b) the facility or activity involved;
   (c) the assumed or established cause and the foreseeable development of the nuclear accident relevant to the transboundary release of the radioactive materials;
   (d) the general characteristics of the radioactive release, including, as far as is practicable and appropriate, the nature, probable physical and chemical form and the quantity, composition and effective height of the radioactive release;
   (e) information on current and forecast meteorological and hydrological conditions, necessary for forecasting the transboundary release of the radioactive materials;
   (f) the results of environmental monitoring relevant to the transboundary release of the radioactive materials;
   (g) the off-site protective measures taken or planned;
   (h) the predicted behaviour over time of the radioactive release.
2. Such information shall be supplemented at appropriate intervals by further relevant information on the development of the emergency situation, including its foreseeable or actual termination.

3. Information received pursuant to sub-paragraph (b) of article 2 may be used without restriction, except when such information is provided in confidence by the notifying State Party.

Article 6

Consultations

A State Party providing information pursuant to sub-paragraph (b) of article 2 shall, as far as is reasonably practicable, respond promptly to a request for further information or consultations sought by an affected State Party with a view to minimizing the radiological consequences in that State.

Article 7

Competent authorities and points of contact

1. Each State Party shall make known to the Agency and to other States Parties, directly or through the Agency, its competent authorities and point of contact responsible for issuing and receiving the notification and information referred to in article 2. Such points of contact and a focal point within the Agency shall be available continuously.
2. Each State Party shall promptly inform the Agency of any changes that may occur in the information referred to in paragraph 1.
3. The Agency shall maintain an up-to-date list of such national authorities and points of contact as well as points of contact of relevant international organizations and shall provide it to States Parties and Member States and to relevant international organizations.

Article 8

Assistance to States Parties

The Agency shall, in accordance with its Statute and

upon a request of a State Party which does not have nuclear activities itself and borders on a State having an active nuclear programme but not Party, conduct investigations into the feasibility and establishment of an appropriate radiation monitoring system in order to facilitate the achievement of the objectives of this Convention.

Article 9

Bilateral and multilateral arrangements

In furtherance of their mutual interests, States Parties may consider, where deemed appropriate, the conclusion of bilateral or multilateral arrangements relating to the subject matter of this Convention.

Article 10

Relationship to other international agreements

This Convention shall not affect the reciprocal rights and obligations of States Parties under existing international agreements which relate to the matters covered by this Convention, or under future international agreements concluded in accordance with the object and purpose of this Convention.

Article 11

Settlement of Disputes

1. In the event of a dispute between States Parties, or between a State Party and the Agency, concerning the interpretation or application of this Convention, the Parties to the dispute shall consult with a view to the settlement of the dispute by negotiation or by any other peaceful means of settling disputes acceptable to them.
2. If a dispute of this character between States Parties cannot be settled within one year from the request for consultation pursuant to paragraph 1, it shall, at

the request of any party to such dispute, be submitted to arbitration or referred to the International Court of Justice for decision. Where a dispute is submitted to arbitration, if, within six months from the date of the request, the parties to the dispute are unable to agree on the organization of the arbitration, a party may the President of the International Court of Justice or the Secretary-General of the United Nations to appoint one or more arbitrators. In cases of conflicting requests by the parties to the dispute, the request to the Secretary-General of the United Nations shall have priority.

3. When signing, ratifying, accepting, approving or acceding to this Convention, a State may declare that it does not consider itself bound by either or both of the dispute settlement procedures provided for in paragraph 2. The other States Parties shall not be bound by a dispute settlement procedure provided for in paragraph 2 with respect to a State Party for which such a declaration is in force.

4. A State Party which has made a declaration in accordance with paragraph 3 may at any time withdraw it by notification to the depositary.

Article 12

Entry into force

1. This Convention shall be open for signature by all States and Namibia, represented by the United Nations Council for Namibia, at the headquarters of the International Atomic Energy Agency in Vienna and at the Headquarters of the United Nations in New York, form 26 September 1986 and 6 October 1986 respectively, until its entry into force or for twelve months, whichever period is longer.

2. A State and Namibia, represented by the United

Nations Council for Namibia, may express its consent to be bound by this Convention either by signature, or by deposit of an instrument of ratification, acceptance or approval following signature made subject to ratification, acceptance or approval, or by deposit of an instrument of accession. The instruments of ratification, acceptance, approval or accession shall be deposited with the depositary.

3. This Convention shall enter into force thirty days after consent to be bound has been expressed by three States.

4. For each State expressing consent to be bound by this Convention after its entry into force, this Convention shall enter into force for that State thirty days after the date of expression of consent.

5. (a) This Convention shall be open for accession, as provided for in this article, by international organizations and regional integrations/organizations constituted by sovereign States, which have competence in respect of the negotiation, conclusion and application of international agreements in matters covered by this Convention.

(b) In matters within their competence such organizations shall, on their own behalf, exercise the rights and fulfil the obligations which this Convention attributes to States Parties.

(c) When depositing its instrument of accession, such an organization shall communicate to the depositary a declaration indicating the extent of its competence in respect of matters covered by this Convention.

(d) Such an organization shall not hold any vote additional to those of its Member States.

Article 13

Provisional application

A State may, upon signature or at any later date before this Convention enters into force for it, declare that it will apply this Convention provisionally.

Article 14

1. A State-Party may propose amendments to this Convention. The proposed amendment shall be submitted to the depositary who shall circulate it immediately to all other States Parties.
2. If a majority of the States Parties request the depositary to convene a conference to consider the proposed amendments, the depositary shall invite all States Parties to attend such a conference to begin not sooner than thirty days after the invitations are issued. Any amendment adopted at the conference by a two-thirds majority of all States parties shall be laid down in a protocol which is open to signature in Vienna and New York by all States Parties.
3. The protocol shall enter into force thirty days after consent to be bound has been expressed by three States. For each State expressing consent to be bound by the protocol after its entry into force, the protocol shall enter into force for that State thirty days after the date of expression of consent.

Article 15

Denunciation

1. A State Party may denounce this Convention by written notification to the depositary.
2. Denunciation shall take effect one year following the date on which the notification is received by the depositary.

Article 16

Depositary

1. The Director General of the Agency shall be the depositary of this Convention.
2. The Director General of The Agency shall promptly notify States Parties and all other States of:
   (a) each signature of this Convention or any protocol of amendment;
   (b) each deposit of an instrument of ratification, acceptance, approval or accession concerning this Convention or any protocol of amendment;
   (c) any declaration or withdrawal thereof in accordance with article 11;
   (d) any declaration of provisional application of this Convention in accordance with article 13;
   (e) the entry into force of this Convention and of any amendment thereto; and
   (f) any denunciation made under article 15.

Article 17

Authentic texts and certified copies

The original of this Convention, of which the Arabic. Chinese, English, French, Russian and Spanish texts are equally authentic, shall be deposited with the Director General of the International Atomic Energy Agency who shall send certified copies to States Parties and all other States.

In witness whereof the undersigned, being duly authorized, have signed this Convention; open for signature as provided for in paragraph 1 of article 12.

Adopted by the General Conference of the International Atomic Energy Agency meeting in special session at Vienna on the twenty-sixth day of September one thousand nine hundred and eighty-six.

4 Convention on Assistance in the Case of a Nuclear Accident or Radiological Emergency (1986)

1. The Convention on Assistance in the Case of a Nuclear Accident or Radiological Emergency was adopted by the General Conference at its special session, 24-26 September 1986, and was opened for signature at Vienna on 26 September 1986 and at New York on 6 October 1986,

2. The text of the Convention, taken from a certified copy, is reproduced herein for the information of all Members.

The date of entry into force will be announced in an Addendum to this document.

Convention on assistance in the case of a nuclear accident or radiological emergency

The states parties to this convention,

Aware that nuclear activities are being carried out in a number of States,

Noting that comprehensive measures have been and are being taken to ensure a high level of safety in nuclear activities, aimed at preventing nuclear accidents and minimizing the consequences of any such accident, should it occur,

Desiring to strengthen further international

cooperation in, the safe development and use of nuclear energy,

Convinced of the need for an international framework which will facilitate the prompt provision of assistance in the event of a nuclear accident or radiological emergency to mitigate it consequences,

Noting the usefulness of bilateral and multilateral arrangements on mutual assistance in this area,

Noting the activities of the International Atomic Energy Agency in developing guidelines for mutual emergency assistance arrangements in connection with a nuclear accident or radiological emergency,

Have agreed as follows:

Article 1

General Provisions

1. The States Parties shall cooperate between themselves and with the International Atomic Energy Agency (hereinafter referred to as the "Agency") in accordance with the provisions of this Convention to facilitate prompt assistance in the event of a nuclear accident or radiological emergency to minimize its consequences and to protect life, property and the environment from the effects of radioactive releases.
2. To facilitate such cooperation State Parties may agree on bilateral or multilateral arrangements or, where appropriate, a combination of these, for preventing or minimizing injury and damage which may result in the event of a nuclear accident or radiological emergency.
3. The States Parties request the Agency, acting within the framework of its Statute, to use its best endeavours in accordance with the provisions of this Convention to promote, facilitate and support the

cooperation between States Parties provided for in this Convention.

Article 2

Provision of Assistance

1. If a State Party needs assistance in the event of a nuclear accident or radiological emergency, whether or not such accident or emergency originates within its territory, jurisdiction or control, it may call for such assistance from any other State Party, directly or through the Agency, and from the Agency, or, where appropriate, from other international intergovernmental organizations (hereinafter referred to as "international organizations")
2. A State Party requesting assistance shall specify the scope and type of assistance required and, where practicable, provide the assisting party with such information as may be necessary for that party to determine the extent to which it is able to meet the request. In the event that it is not practicable for the requesting State Party to specify the scope and type of assistance required, the requesting State Party and the assisting party shall, in consultation, decide upon the scope and type of assistance required.
3. Each State Party to which a request for such assistance is directed shall promptly decide and notify the requesting State Party, directly or through the Agency, whether it is in a position to render the assistance requested, and the scope and terms of the assistance that might be rendered.
4. States Parties shall, within the limits of their capabilities, identify and notify the Agency of experts, equipment and materials which could be made available for the provision of assistance to other States Parties in the event of a nuclear accident

or radiological emergency as well as the terms, especially financial, under which such assistance could be provided.

5. Any State Party may request assistance relating to medical treatment or temporary relocation into the territory of another State Party of people involved in a nuclear accident or radiological emergency.
6. The Agency shall respond, in accordance with its Statute and as provided for in this Convention, to a requesting State Party's or a Member State's request for assistance in the event of a nuclear accident or radiological emergency by:
   (a) making available appropriate resources allocated for this purpose;
   (b) transmitting promptly the request to other States and international organizations which, according to the Agency's information, may possess the necessary resources; and
   (c) if so requested by the requesting States, coordinating the assistance at the international level which may thus become available.

Article 3

Direction and Control of Assistance

Unless otherwise agreed:

(a) the overall direction, control, co-ordination and supervision of the assistance shall be the responsibility within its territory of the requesting State. The assisting party should, where the assistance involves personnel, designate in consultation with the requesting State, the person who should be in charge of and retain immediate operational supervision over the personnel and the equipment provided by it. The designated person should exercise such supervision in cooperation with the appropriate authorities of the requesting State;

(b) the requesting State shall provide, to the extent of its capabilities, local facilities and services for the proper and effective administration of the assistance. It shall also ensure the protection of personnel, equipment and materials brought into its territory by or on behalf of the assisting party for such purpose;

(c) ownership of equipment and materials provided by either party during the periods of assistance shall be unaffected, and their return shall be ensured;

(d) a State Party providing assistance in response to a request under paragraph 5 of article 2 shall co-ordinate that assistance within its territory.

Article 4

Competent authorities and points of contact

1. Each State Party shall make known to the Agency and to other States Parties, directly or through the Agency, its competent authorities and point of contact authorized to make and receive requests for and to accept offers of assistance. Such points of contact and a focal point within the Agency shall be available continuously.
2. Each State Party shall promptly inform the Agency of any change that may occur in the information referred to in paragraph 1.
3. The Agency shall regularly and expeditiously provide to State Parties, Member States and relevant international organizations the information referred to in paragraphs 1 and 2.

Article 5

Functions of the Agency

The States Parties request the Agency, in accordance with

paragraph 3 of article 1 and without prejudice to other provisions of this Convention, to:

(a) collect and disseminate to State Parties and Members information concerning:

 (i) experts, equipment and materials which could be made available in the event of nuclear accidents or radiological emergencies;

 (ii) methodologies, techniques and available result of research relating to response to nuclear accidents or radiological emergencies;

(b) assist a State Party or a Member State when requested in any of the following or other appropriate matters:

 (i) preparing both emergency plans in the case of nuclear accidents and radiological emergencies and the appropriate legislation;

 (ii) developing appropriate training programmes for personnel to deal with nuclear accidents and radiological emergencies;

 (iii) transmitting requests for assistance and relevant information in the event of a nuclear accident or radiological emergency;

 (iv) developing appropriate radiation monitoring programmes, procedures and standards;

 (v) conducting investigations into the establishing appropriate radiation monitoring systems;

(c) make available to a State Party or a Member State requesting assistance in the event of nuclear accident or radiological emergency appropriate resources allocated for the purpose of conducting an initial assessment of the accident or emergency;

(d) offer its good offices to the States Parties and Members States in the event of a nuclear accident or radiological emergency;

(e) establish and maintain liaison with relevant international organization for the purposes of obtaining and exchanging relevant information and data, and make a list of such organizations available to States Parties, Member States and the aforementioned organizations.

Article 6

Confidentiality and Public Statements

1. The requesting State and the assisting party shall protect the confidentiality of any confidential information that becomes available to either of them in connection with the assistance in the event of a nuclear accident or radiological emergency. Such information shall be used exclusively for the purpose of the assistance agreed upon.
2. The assisting party shall make every effort to coordinate with the requesting State before releasing information to the public on the assistance provided in connection with a nuclear accident or radiological emergency.

Article 7

Reimbursement of Costs

1. An assisting party may offer assistance without costs to the requesting State. When considering whether of offer assistance on such a basis, the assisting party shall take into account:
   (a) the nature of the nuclear accident or radiological emergency;
   (b) the place or origin of the nuclear accident or radiological emergency;
   (c) the needs of developing countries;
   (d) the particular needs of countries without nuclear facilities; and

(e) any other relevant factors.

2. When assistance is provided wholly or partly on a reimbursement basis, the requesting State shall reimburse the assisting party for the costs incurred for the services rendered by persons or organizations acting on its behalf, and for all expenses in connection with the assistance to the extent that such expenses are not directly defrayed by the requesting State. Unless otherwise agreed, reimbursement shall be provided promptly after the assisting party has presented its request for reimbursement to the requesting State, and in respect of costs other than local costs, shall be freely transferrable.

3. Notwithstanding paragraph 2, the assisting party may at any time waive, or agree to the postponement of, the reimbursement in whole or in part. In considering such waiver or postponement, assisting parties shall give due consideration to the needs of developing countries.

Article 8

Privileges, Immunities and Facilities

1. The requesting State shall afford to personnel of the assisting party and personnel acting on its behalf the necessary privilege, immunities and facilities for the performance of their assistance functions.

2. The requesting State shall afford the following privilege and immunities to personnel of the assisting party or personnel acting on its behalf who have been duly notified to and accepted by the requesting State:

   (a) immunity from arrest, detention and legal process, including criminal, civil and administrative jurisdiction, of the requesting

State, in respect of acts or omissions in the performance of their duties; and

(b) exemption from taxation, duties or other charges, except those which are normally incorporated in the price of goods or paid for services rendered, in respect of the performance of their assistance functions.

3. The requesting State shall:

(a) afford the assisting party exemption from taxation, duties or other charges on the equipment and property brought into the territory of the requesting State by the assisting party for the purpose of the assistance; and

(b) provide immunity from seizure, attachment or requisition of such equipment and property.

4. The requesting State shall ensure the return of such equipment and property. If requested by the assisting party, the requesting State shall arrange, to the extent it is able to do so, for the necessary decontamination of recoverable equipment involved in the assistance before its return.

5. The requesting State shall facilitate the entry into, stay in and departure from its national territory of personnel notified pursuant to paragraph 2 and of equipment and property involved in the assistance.

6. Nothing in this article shall require the requesting State to provide its nationals or permanent residents with the privileges and immunities provided for in the foregoing, paragraphs.

7. Without prejudice to the privileges and immunities, all beneficiaries enjoying such privileges and immunities under this article have a duty to respect the laws and regulations of the requesting State. They shall also have the duty not to interfere in the domestic affairs of the requesting State.

8. Nothing in this article shall prejudice rights and obligations with respect to privileges and immunities afforded pursuant to other international agreements or the rules of customary international law.
9. When signing, ratifying, accepting, approving or acceding to this Convention, a State may declare that it does not consider itself bound in whole or in part by paragraphs 2 and 3.
10. A State Party which has made a declaration in accordance with paragraph 9 may at any time withdraw it by notification to the depositary.

Article 9

Transit of personnel, equipment and property

Each State Party shall, at the request of the requesting State or the assisting party, seek to facilitate the transit through its territory of duly notified personnel, equipment and property involved in the assistance to and from the requesting State.

Article 10

Claims and compensation

1. The States Parties shall closely cooperate in order to facilitate the settlement of legal proceedings and claims under this article.
2. Unless otherwise agreed, a requesting State shall in respect of death or of injury to persons, damage to or loss of property, or damage to the environment caused within its territory or other area under its jurisdiction or control in the course of providing the assistance requested:
   (a) not being any legal proceedings against the assisting party or persons or other legal entities acting on its behalf,

(b) assume responsibility for dealing with legal proceedings and claims brought by third parties against the assisting party or against persons or other legal entities acting on its behalf;

(c) hold the assisting party or persons or other legal entities acting on its behalf harmless in respect of legal proceedings and claims referred to in sub-paragraph (b); and

(d) compensate the assisting party or persons or other legal entities acting on its behalf for;

(i) death of or injury to personnel of the assisting party or persons acting on its behalf for:

(ii) loss of or damage to non-consumable equipment or materials related to the assistance;

except in cases of wilful misconduct by the individuals who caused the death, injury, loss or damage.

3. This article shall not prevent compensation or indemnity available under any applicable international agreement or national law of any State.

4. Nothing in this article shall require the requesting State to apply paragraph 2 in whole or in part to its nationals or permanent residents.

5. When signing, ratifying, accepting, approving or acceding to this Convention, a State may declare:

(a) that it does not consider itself bound in whole or in part by paragraph 2;

(b) that it will not apply paragraph 2 in whole or in part in cases of gross negligence by the individuals who caused the death, injury, or damage.

6. A State Party which has made a declaration in accordance with paragraph 5 may at any time withdraw it by notification to the depositary.

Article 11

Termination of assistance

The requesting State or the assisting party may at any time, after appropriate consultations and by notification in writing, request the termination of assistance received or provided under this Convention. One such a request has been made, the parties involved shall consult with each other to make arrangements for the proper conclusion of the assistance.

Article 12

Relationship to other international agreements

This Convention shall not affect the reciprocal rights and obligations of States Parties under existing international agreements which relate to the matter covered by this Convention, or under future international agreements concluded in accordance with the object and purpose of this Convention.

Article 13

Settlement of disputes

1. In the event of a dispute between State Parties, or between a State Party and the Agency, concerning the interpretation or application of this Convention, the parties to the dispute shall consult with a view to the settlement of the dispute by negotiation or by any other peaceful means of settling disputes acceptable to them.
2. If a dispute of this character between States Parties cannot be settle within one year from the request for consultation pursuant to paragraph 1, it shall, at the request of any party to such dispute, be submitted to Arbitration or referred to the International Court of Justice for decision. Where a dispute is submitted to arbitration, if, within six months from the date of the

request, the parties to the dispute are unable to agree on the organization of the arbitration, a party may request the President of the International Court of Justice or the Secretary-General of the United Nations to appoint one or more arbitrators. In cases of conflicting requests by the parties to the dispute, the request to the Secretary-General of the United Nations shall have priority.

3. When signing, ratifying, accepting, approving or acceding to this Convention, a State may declare that it does not consider itself bound by either or both of the dispute settlement procedures provided for in paragraph 2. The other States Parties shall not be bound by a dispute settlement procedure provided for in paragraph 2 with respect to a Party for which such a declaration is in force.

4. A State Party which has made a declaration in accordance with paragraph 3 may at any time withdraw it by notification to the depositary.

Article 14

Entry into force

1. This Convention shall be open for signature by all States and Namibia, represented by the United Nations Council for Namibia, at the Headquarters of the International Atomic Energy Agency in Vienna and at the Headquarter of the United Nations in New York, from 26 September 1986 and 6 October 1986 respectively, until its entry into force or for twelve months, whichever period is longer.

2. A State and Namibia, represented by the United Nations Council for Namibia, may express its consent to be bound by this Convention either by signature, or by deposit of an instrument of ratification, acceptance or approval following

signature made subject to ratification, acceptance or approval, or by deposit of an instrument of accession. The instruments of ratification, acceptance, approval or accession shall be deposited with the depositary.

3. This Convention shall enter into force thirty days after consent to be bound has been expressed by three States.

4. For each State expressing consent to be bound by this Convention after its entry into force, this Convention shall enter into force for that State thirty days after the date of expression of consent.

5. a) This Convention shall be open for accession, as provided for in this article, by international organizations and regional integration organizations constituted by sovereign States, which have competence in respect of the negotiation, conclusion and application of international agreements in matters covered by this Convention.

(b) In matters within their competence such organizations shall, on their own behalf, exercise the rights and fulfil the obligations which this Convention attributes to States Parties.

(c) When depositing its instrument of accession, such an organization shall communicate to the depositary a declaration indicating the extent of its competence in respect of matters covered by this Convention.

(d) Such an organization shall not hold any vote additional to those of its Member States.

Article 15

Provisional application

A State may, upon signature or at any later date before this Convention enters into force for it, declare that it will apply this Convention provisionally.

Article 16

Amendments

1. A State Party may propose amendments to this Convention. The proposed amendment shall be submitted to the depositary who shall circulate it immediately to all other States Parties.
2. If a majority of the State Parties request the depositary to convene a conference to consider the proposed amendments, the depositary shall invite all States Parties to attend such a conference to begin not sooner than thirty days after the invitations are issued. Any amendment adopted at the conference by a two-thirds majority of all States Parties shall be laid down in a protocol which is open to signature in Vienna and New York by all States Parties.
3. The protocol shall enter into force thirty days after consent to be bound has been expressed by three States. For each State expressing consent to be bound by the protocol after its entry into force, the protocol shall enter into force for that State thirty days after the date of expression of consent.

Article 17

Denunciation

1. A State Party may denounce this Convention by written notification to the depositary.
2. Denunciation shall take effect one year following the date on which the notification is received by the depositary.

Article 18

Depositary

1. The Director General of the Agency shall be the depositary of this Convention.
2. The Director General of the Agency shall promptly notify States Parties and all other States of:
   (a) each signature of this Convention or any protocol of amendment;
   (b) each deposit of an instrument of ratification, acceptance, approval or accession concerning this Convention or any protocol of amendment;
   (c) any declaration or withdrawal thereof in accordance with articles 8, 10 and 13;
   (d) any declaration of provisional application of this Convention in accordance with article 15;
   (e) the entry into force of this Convention and of any amendment thereto; and
   (f) any denunciation made under article 17.

Article 19

Authentic texts and certified copies

The original of this Convention, of which the Arabic, Chinese, English, French, Russian and Spanish texts are equally authentic, shall be deposited with the Director General of the International Atomic Energy Agency who shall send certified copies to States Parties and all other States.

In witness whereof the undersigned, being duly authorized, have singed this Convention, open for signature as provided for in paragraph 1 of article 14.

Adopted by the General Conference of the International Atomic Energy Agency meeting in special session at Vienna on the twenty-sixth day of September one thousand nine hundred and eighty-six.

5 Status of Convention on Early Notifications of Nuclear Accident

Reservations/Declarations made upon or following expressing consent to be bound and objections thereto

Argentina [17 January 1990]

Convention on Early Notification of a Nuclear Accident

"In accordance with Article 11, paragraph 3, the Argentine Republic does not consider itself bound by any of the dispute settlement procedures provided for in Article 11, paragraph 2 of the Convention."

Convention on Assistance in the Case of a Nuclear Accident or Radiological Emergency

"In accordance with Article 8, paragraph 9, the Argentine Republic does not consider itself bound by any of the provisions concerning privileges and immunities under Article 8, paragraphs 2 and 3 of the Convention;

"In accordance with Article 10, paragraph 5, the Argentine Republic does not consider itself bound by any of the provisions concerning claims and compensation under Article 10, paragraph 2;

"In accordance with Article 13, paragraph 3, the Argentine Republic does not consider itself bound by any of the dispute settlement procedures provided for in Article 13, paragraph 2 of the Convention."

(Original in Spanish; translation by the Secretariat)

Australia [22 September 1987]

Convention on Assistance in the Case of a Nuclear Accident or Radiological Emergency

"Declaring as permitted under Article 8.9 of the Convention, that Australia will not be bound by Articles 8.2 and 8.3."

(Original in English)

Austria [21 November 1989]

Convention on Assistance in the Case of a Nuclear Accident or Radiological Emergency

"I have the honour to inform you in conformity with article 10 paragraph 5(b) of the Convention of Assistance in the Case of a Nuclear Accident or Radiological Emergency that Austria will not apply paragraph 2 of the aforementioned article in cases of gross negligence by the individuals who caused death, injury, loss or damage."

(Original in English)

Belarus [26 January 1987]

The following identical reservation was received in respect of both Conventions:

"The Byelorussian SSR will not consider itself bound by the provisions of Article II, paragraph 2 of the Convention on Early Notification of a Nuclear Accident and Article 13, paragraph 2 of the Convention on Assistance in the Case of a Nuclear Accident or Radiological Emergency, which envisage the possibility of submitting a dispute between States Parties to arbitration or referring it to the International Court of Justice at the request of any party and States that, for submission of any international dispute to arbitration or referral to the international Court of Justice, the agreement of all parties in each individual case is necessary."

(Original in Russian; translation by the Secretariat)

Bulgaria [24 February 1988]

Convention on Early Notification of a Nuclear Accident

"The people's Republic of Bulgaria does not consider itself bound by the provisions of article 11, paragraph 2, which provide the possibility for submission of the disputes to arbitration or to the International Court of Justice at the request of any party to such dispute and declares that for the submission of any international dispute to arbitration or to the International Court of Justice, the consent of all parties concerned in each individual case is necessary."

Convention on Assistance in the Case of a Nuclear Accident or Radiological Emergency

"The People's Republic of Bulgaria does not consider itself bound by the provisions of article 13, paragraph 2, which provide the possibility for submission of the disputes to arbitration or to the International Court of Justice at the request of any party to such dispute and declares that for the submission of any international dispute to arbitration or to the International Court of Justice, the Consent of all parties concerned in each individual case is necessary."

(Original in English and Bulgarian; supplied by the Government)

China [10 September 1987]

Convention on Early Notification of a Nuclear Accident

"China shall not be bound by the two dispute settlement procedures provided for in paragraph 2, Article 11 of the Convention of Early Notification of a Nuclear Accident."

Convention on Assistance in the Case of a Nuclear Accident or Radiological Emergency

"China shall not apply paragraph 2, Article 10 of the Convention on Assistance in the Case of a Nuclear Accident or Radiological Emergency in cases of gross negligence by the individuals who caused the death, injury, loss or damage.

"China shall not be bound by the tow dispute settlement procedures provided for in paragraph 2, Article 13."

(Original in Chinese and English; supplied by the Government).

Cuba [8 January 1991]

Convention on Early Notification of a Nuclear Accident

"The Government of the Republic of Cuba declares, in accordance with paragraph 3 of Article 11 of the Convention on Early Notification of a Nuclear Accident, that it does not consider itself bound by the procedure stipulated in paragraph 2 for the settlement of disputes."

Nuclear Accident or Radiological Emergency

"The Government of the Republic of Cuba declares, in accordance with paragraph 3 of article 13 of the Convention on Assistance in the Case of a Nuclear Accident or Radiological Emergency, that it does not consider itself bound by the procedure stipulated in paragraph 2 for the settlement of disputes."

(Original in Spanish; translation by the secretariat)

Egypt [6 July 1988]

Convention on Early Notification of a Nuclear Accident

"1. The Arab Republic of Egypt views Articles 1 and 2 of the Convention dealing with the scope of application thereof in the light of the official declarations made by the representatives of China,

France, the Soviet Union, the United Kingdom and the United States of America as to the readiness of their Governments on a Voluntary basis to notify the International Atomic Energy Agency and any other States affected by any accident which is not specified in Article 1 of the Convention and which may have transboundary radiological consequences.

"2. The Arab Republic of Egypt declares that it does not consider itself bound by any of the procedures for settlement of disputes referred to in Article 11, paragraph 2."

(Original in Arabic; translation by the Secretariat)

Egypt [17 October 1988]

Convention on Assistance in the Case of a Nuclear Accident or Radiological Emergency

"1. The Arab Republic of Egypt considers that Article 5 of the convention relating to the "Functions of the Agency" Shall be read and applied in the light of, and in accordance with, Article 2.6;

"2. The Arab Republic of Egypt interprets Article 7 to mean that the needs of developing countries shall be taken specially into account during consideration of requests for assistance in the event of nuclear accidents;

"3. The Arab Republic of Egypt considers that the obligations relating to privileges and immunities referred to in Article 8 shall be applied in accordance with Egyptian law;

"4. The Arab Republic of Egypt declares that it does not consider itself bound by any of the dispute settlement procedures provided for in Article 13, paragraph 2."

(Original in Arabic; translation by the Secretariat)

Finland [27 November 1990]

Convention on Assistance in the Case of a Nuclear Accident or Radiological Emergency

"Finland will not apply paragraph 2 of Article 10 in cases of gross negligence by the individuals who caused the death, injury, loss or damage."

(Original in English)

Food and Agriculture Organization

[19 October 1990]

Convention on Early Notification of a Nuclear Accident

"Pursuant to Article 12, paragraph 5(c), the Director General of FAO declares that, within its constitutional mandate to monitor and evaluate the world food security situation, the Food and Agriculture Organization of the United Nations is competent to assess the qualitative and quantitative effects of all contaminants including radionuclides on food supplies, and to advise governments on acceptable levels of radionuclides appearing in agricultural, fisheries and forestry products entering national and international trade."

Convention on Assistance in the Case of a Nuclear Accident or Radiological Emergency

"Pursuant to Article 14, paragraph 5 (c), the Director-General of FAO declares that, within its constitutional mandate to monitor and evaluate the world food security situation, the Food and Agriculture Organization of the United Nations is competent to advise governments on measures to be taken in terms of the agricultural, fisheries and forestry practices to minimize the impact of radionuclides and to develop emergency procedures for alternative agricultural practices and for decontamination of agricultural, fisheries and forestry products, soil and water."

(Original in English)

France [6 March 1989]

Convention on Early Notification of a Nuclear Accident

"The Government of the French Republic declares, in accordance with Article 11.3, that France does not consider itself bound by the provisions of paragraph 2 of that Article."

Convention on Assistance in the Case of a Nuclear Accident or Radiological Emergency

"The Government of the French Republic declares, in accordance with Article 8.9 that France does not consider itself bound by the provisions of paragraphs 2 and 3 of that Article."

"The Government of the French Republic declares, in accordance with Article 10.5, that France does not consider itself bound by paragraph 2 of that Article;

"The Government of the French Republic declares, in accordance with Article 13.3, that France does not consider itself bound by the provisions of paragraph 2 of that Article."

(Original in French; translation by the Secretariat)

Germany, Fed, Rep Of [14 September 1989]

The following identical declaration was received with regard to the Convention on Early Notification of a Nuclear Accident and the Convention on Assistance in the Case of a Nuclear Accident or Radiological Emergency: "The said Convention shall also apply to Land Berlin with effect from the date on which it has entered into force for the Federal Republic of Germany."

(Original in English)

Hungary [10 March 1987]

(The following reservations were subsequently withdrawn 30 Nov. 1989)

Convention on Early Notification of a Nuclear Accident

"The Hungarian People's Republic does not consider itself bound by the dispute settlement procedures provided for in paragraph 2 of article II of the Convention since, in its opinion, the jurisdiction of any arbitral tribunal or of the International Court of Justice can be founded only on the voluntary prior acceptance of such jurisdiction by all the Parties concerned."

Convention on Assistance in the case of a Nuclear Accident or Radiological Emergency

"The Hungarian People's Republic does not consider itself bound by the dispute settlement procedures provided for in article 13, paragraph 2, of the Convention since, in its opinion, the jurisdiction of any arbitral tribunal or of the International Court of Justice can be founded only on the voluntary prior acceptance of such jurisdiction by all the Parties concerned."

(Original in English and Hungarian; supplied by the Government)

India [28 January 1988]

Convention on Early Notification of a Nuclear Accident

"1. The Government of India considers that the convention suffers from serious and inherent defects in as much as it differentiates between nuclear weapon states and non-nuclear weapon states. The Convention is defective as it does not contain a legal provision to make it mandatory on the nuclear weapon States to notify accidents involving nuclear weapons or weapon tests. The Government of India feels that the Convention should have provided for notification of nuclear accidents in any nuclear facility, vessel, aircraft, spacecraft, etc. used for peaceful or military purposes as well as nuclear weapons.

"2. The Government of India is disappointed at the outcome of the Convention because it does pot cover all accidents. It should have been a comprehensive Convention covering accidents from whatever source-civil or military, including accidents emanating from nuclear weapons or nuclear weapon tests, since the transboundary effects of radiological safety significance from any source whatsoever, would be equally damaging. Nevertheless, the Government of India has ratified the Convention, in view of the solemn assurances that has been given by the five nuclear weapon States to the effect that they undertake to notify all accidents. This is in keeping with our policy of according to public, declarations of state policy equal validity with other international commitments.

"3. The Government of India hereby declares that it does not consider itself bound by the dispute settlement procedures provided for in para 2 of Article 11."

Convention on Assistance in the Case of a Nuclear Accident or Radiological Emergency

"1. The Government of India hereby declares that it does not consider itself bound by paragraphs 2 and 3 of Article 8 of the Convention.

"2. The Government of India hereby declares that it does not consider itself bound by para 2 of Article 10 of the Convention.

"3. The Government of India hereby declares that it does not consider itself bound by the dispute settlement procedures provided for in para 2 of Article 13 of the Convention."

(Original in English)

Iraq [21 July 1988]

Convention on Early Notification of a Nuclear Accident)

1. Iraq does not consider itself bound by* "The provision contained in Article 11, paragraph 2 of the Convention concerning the obligation to accept arbitrators appointed by the president of the International Court of Justice or the Secretary-General of the United Nations.

"2. This ratification does not in any way imply recognition of Israel or entering into any relationship with it."

(Original in Arabic "*preceding words added by translator)

Convention on Assistance in the Case of a Nuclear Accident of Radiological Emergency

"1. Article 8 Concerning immunity for legal proceedings: availing ourselves of the powers granted to the States signing the Convention under Article 10, paragraphs 5 (a) and (b), we consider that cases of gross negligence should be excluded from absolute immunity so that the assisting party will not be exempt from responsibility."

2. Iraq does not consider itself bound by Article 13, Paragraph 2 concerning the obligation to accept arbitrators appointed by the president of the International Court of Justice or the Secretary-General of the United Nations.

3. This ratification does not in any way imply recognition of Israel or entering into any relationship with it."

(Original in Arabic; Preceding words added by translator)

Israel [4 January 1989]

Objection to reservations made by the Republic of Iraq upon ratification of the Convention on Early

Notification of a Nuclear Accident and the Convention on Assistance in the Case of a Nuclear Accident or Radiological Emergency

"The Government of the State of Israel has noted that the instruments of Ratification of the Republic of Iraq of the above mentioned Conventions contain a declaration in respect of Israel. In the view of the Government of the State of Israel, such declaration, which is explicitly of a political character, is incompatible with the purposes and objectives of these Conventions and cannot in any way affect whatever obligations are binding upon Iraq under general international law or under particular Conventions.

"The Government of the State of Israel will, in so far as concerns the substance of the matter, adopt towards the Republic of Iraq an attitude of complete reciprocity." (Original in English)

Israel [25 May 1989]

Convention on Early Notification of a Nuclear Accident

"The Government of the State of Israel declares, in accordance with Article 11.3, that Israel does not consider itself bound by the provisions of paragraph 2 of that Article."

Convention on Assistance in the Case of a Nuclear Accident or Radiological Emergency

"The Government of the State of Israel declares that Israel does not consider itself bound by the provisions of:

paragraph 2 (a) of Article 8

paragraph 2 of Article 10

paragraph 2 of Article 13."

(Original in English)

Israel [19 January 1990]

Objection to a reservation made by Saudi Arabia upon

accession to the Convention on Early Notification of a Nuclear Accident and the Convention on Assistance in the Case of a Nuclear Accident or Radiological Emergency.

"The Government of the State of Israel has noted that the instrument of accession of Saudi Arabia to the above mentioned Conventions contains a reservation in respect of Israel. In the view of the Government of the State of Israel, such reservation, which is explicitly of a political character, is incompatible with the purposes and objectives of these Conventions and cannot in any way affect whatever obligations are binding upon Saudi Arabia under general international law or under particular Conventions.

"The Government of the State of Israel will, in so far as concerns the substance of the matter, adopt towards Saudi Arabia an attitude of complete reciprocity."

(Original in English)

Italy [8 February 1990]

Convention on Early Notification of a Nuclear Accident

"The Italian Government declares that the clauses of article 1 are not satisfactory, in so far as they impose on a contracting party the obligation to notify only accident from which derive the release of radioactive materials which might transcend or has transcended an international boundary, or could have other consequences outside its jurisdiction or control.

"The Italian Government considers that every accident should be notified, also those which have consequences limited to the territory of the State concerned."

(Original in English)

Italy [25 October 1990]

Convention on Assistance in the Case of a Nuclear Accident or Radiological Emergency

"In accordance with Article 8.9, the Government of the Italian Republic expresses the following reservation.

"Italy understand the term 'duties' in Article 8.2 (b) to refer only to customs duties Moreover, it specifies that exemption from taxation, duties or other charges cannot apply to value-added tax (VAT) and that in no case can said exemptions apply to Italian nationals or persons resident in Italy on a permanent basis.

"Pursuant to Article 10.5 (b), the Government of the Italian Republic declares that it does not consider itself bound by paragraph 2 in cases of gross negligence by the individuals who caused the death, injury, loss or damage.

"Lastly,·the Government of the Italian Republic makes the following interpretative declaration:

(a) The general provision contained in Article 8.1 shall concern only those privileges, immunities and exemptions specified in the following paragraphs, to the exclusion of all else;

(b) The immunity referred to in Article 8.2 (a) shall be understood to the affordable in respect of acts or omissions committed in the performance of and in connection with the duties discharged."

(Original in French; translation by the secretariat)

Japan[9 June 1987]

Convention on Assistance in the Case of a Nuclear Accident or Radiological Emergency

"The Government of Japan declares that it does not consider itself bound by paragraph 2(b) of Article 8 with respect to the income tax, local inhabitant taxes and the enterprise tax as well as any identical or substantially similar taxes on personnel acting on behalf of an assisting party and that it will afford to the said personnel exemption from these taxes to the extent provided for in a convention for the avoidance of double taxation between Japan and the State of which the personnel is a resident."

(Original in English and Japanese; supplied by the Government)

Korea, Rep. Of [8 June 1990]

Convention on Assistance in the Case of a Nuclear Accident or Radiological Emergency

"The Government of the Republic of Korea declares pursuant to paragraph 9 of Article 8 of the Convention on Assistance in the Case of a Nuclear Accident or Radiological Emergency that it does not consider itself bound by paragraphs 2 and 3 of that Article. And the Government of the Republic of Korea declares pursuant to paragraph 5 of Article 10 of the said Convention that it does not consider itself bound by paragraph 2 of that Article."

(Original in English and Korean; supplied by the Government)

Malaysia [September 1987]

Convention on Early Notification of a Nuclear Accident

"The Government of Malaysia declares, in accordance with paragraph 3 of Article 11, that Malaysia does not consider itself bound by the dispute settlement procedures provided for in paragraph 2 of that Article.

Convention on Assistance in the Case of a Nuclear Accident of Radiological Emergency

"The Government of Malaysia declares, in accordance with paragraph 3 of Article 13, that Malaysia does not consider itself bound by the dispute settlement procedures provided for in paragraph 2 of that Article."

(Original in English)

Mauritius [17 August 1992]

Convention on Early Notification of a Nuclear Accident

"(a) The Government of the Republic of Mauritius regrets that the scope of the Convention on Early Notification of a Nuclear Accident does not also cover nuclear emergencies resulting from military activities which involve nuclear weapons since the potential transboundary radiological effect would be equally injurious;

"(b) in accordance with article (ii) paragraph (3) of the Convention on early Notification of a Nuclear Accident, the Republic of Mauritius does not consider itself bound by the provisions of paragraph (2) article (11) thereof."

Convention on Assistance in the Case of a Nuclear Accident or Radiological Emergency

"In accordance with Article 13 paragraph 3 of the Convention Mauritius declares that it does not consider itself bound by any of the dispute settlement procedures set out in paragraph 2 of Article 13."

(Original in English)

Monaco [19 July 1989]

Convention on Early Notification of a Nuclear Accident

"The Principality of Monaco declares, in conformity, with paragraph 3 of Article 11, that it does not consider itself bound by the provisions of paragraph 2 of that Article."

Convention on Assistance in the Case of a Nuclear Accident or Radiological Emergency

"The Principality of Monaco declares that:

(1) In conformity with paragraph 9 of Article 8, it does not consider itself bound by the provisions of paragraphs 2 and 3 of that Article;

(2) In conformity with paragraph 5 of Article 10, it does not consider itself bound by the provisions of paragraph 2 of that Article;

(3) In conformity with paragraph 3 of Article 13, it does not consider itself bound by the provisions of paragraph 2 of that Article."

(Original in French)

Mongolia [11 June 1987]

(The following reservations were subsequently withdrawn 18 June 1990)

Convention on Early Notification of a Nuclear Accident

"The Mongolian People's Republic States that it does not consider itself bound by the provisions of Article 11, paragraph 2 of the Convention on Early Notification of a Nuclear Accident, concerning the procedure of the settlement of disputes arising from the interpretation or application of the Convention. In its opinion, for submission of any dispute of such nature to arbitration or the International Court of Justice the consent of all the parties to the dispute is necessary."

Convention on Assistance in the Case of a Nuclear Accident or Radiological Emergency

"The Mongolian People's Republic States that it does not consider itself bound by the provisions of Article 13, paragraph 2 of the Convention on Assistance in the Case of a Nuclear Accident or Radiological Emergency, concerning the procedure of the settlement of disputes arising from the interpretation or' application of the Convention. In its opinion, for submission of any dispute of such nature to arbitration or the International Court of Justice the consent of all the parties to the dispute is necessary."

(Original in English and Mongolian; supplied by the Government)

New Zealand [11 March 1989]

Convention on Assistance in the Case of a Nuclear Accident or Radiological Emergency

"In accordance with Article 8 (9) of that Convention I declare on behalf of the Government of New Zealand that New Zealand does not consider itself bound by the provisions of Article 8 (2) (a) and Article 8 (3) (b) of the Convention."

(Original in English)

Pakistan [11 September 1989]

Convention on Early Notification of a Nuclear Accident

"The Islamic Republic of Pakistan does not consider itself bound by the provisions of Article 11, paragraph 2, which provides the possibility for submission of disputes to arbitration or to the International Court of Justice at the request of any party to such dispute and declares that for the submission of any international dispute to arbitration or to the International Court of Justice, the consent of all parties concerned in each individual case is necessary."

Convention on Assistance in the Case of a Nuclear Accident or Radiological Emergency

"The Government of Islamic Republic of Pakistan declares that it does not consider itself bound by paragraphs 2 and 3 of Article 8.

"The Islamic Republic of Pakistan does not consider itself bound by the provisions of Article 10, paragraph 2, with respect to cases of gross negligence by the individuals who caused the death, injury, loss or damage.

"The Islamic Republic of Pakistan does not consider itself bound by the provisions of Article 13, paragraph 2, which provides the possibility for submission of disputes to arbitration or to the International Court of Justice at the request of any party to such dispute and declares that for the submission of any 'international dispute to arbitration or to the International Court of Justice, the consent of all parties concerned in each individual case is necessary."

(Original in English)

Poland [24 March 1988]

Convention on Early Notification of a Nuclear Accident

"......the Polish People's Republic does not consider itself bound by the provisions of paragraph 2 of article 11 of the Convention."

Convention on Assistance in the case of a Nuclear Accident or Radiological Emergency

"......the polish People's Republic does not consider itself bound by the provisions of paragraph 2 of article 13 of the Convention."

(Original in English and Polish; supplied by the Government)

Romania [12 June 1990]

Convention of Early Notification of a Nuclear Accident
Convention on Assistance in the Case of a Nuclear Accident or Radiological Emergency

"Romania does not consider itself bound by the provisions of Article 11, paragraph 2 of the Convention on Early Notification of a Nuclear Accident and article 13, paragraph 2 of the Convention on Assistance in the Case of a Nuclear Accident or Radiological Emergency and declares that the submission of any international dispute on the interpretations or the application of those Conventions to arbitration or its referral to the International Court of Justice requires the agreement of all the parties of the dispute."

(Original in Romanian and French; translation by the Secretariat)

Russian Federation [23 December 1986]

The following identical reservation was received in respect of both conventions; "The USSR will not consider itself bound by the provisions of article II, paragraph 2 of the Convention on Early Notification of a Nuclear Accident and article 13, paragraph 2 of the Convention on assistance in the Case of a Nuclear Accident or Radiological Emergency, which envisage the possibility of submission of any international dispute to arbitration or referring it to the International Court of Justice at the request of any party, and states that for the submission of any international dispute to arbitration or referral to the International Court of Justice the agreement of all parties in each individual case is necessary."

(Original in Russian; translation by the Secretariat)

Saudi Arabia [3 November 1989]

"1. Convention on Early Notification of a Nuclear Accident

(1) The Government of the Kingdom of Saudi Arabia, declares that the provisions of Article 1 are unsatisfactory in that they impose on the States Parties the obligation to notify only those accidents resulting in a release of radioactive material which has crossed or may cross an international boundary or those which may have consequences outside their jurisdiction control. The Government of the Kingdom of Saudi Arabia considers that all accidents should be notified, including those with consequences limited to the territory of the State concerned, regardless of the source of the accident, whether civil or military, including accidents resulting from nuclear weapons or nuclear-weapons tests, since transboundary effects from any source which have safety significance may cause harm to all without distinction.

"(2) In accordance with Paragraph 3 of Article 11 the Government of the Kingdom of Saudi Arabia declares that it does not consider itself bound by any of the dispute settlement procedures provided for in paragraph 2 of that Article."

"II. Convention on Assistance in the Case of a Nuclear Accident or Radiological Emergency

"(1) The Government of the Kingdom of Saudi Arabia will not be bound in whole or in part by paragraphs 2 and 3 of Article 8 relating to the grant of privileges, immunities and facilities to the assisting parties.

"(2) The Government of the Kingdom of Saudi Arabia does not consider itself bound in whole or in part by paragraph 2 of Article 10 relating to claims and compensation. The Government of the Kingdom of Saudi Arabia will apply local law in deciding on the relevant procedures.

"(3) In accordance with paragraph 3 of Article 13, the Government of the Kingdom of Saudi Arabia declares that it does not consider itself bound by any of the dispute settlement procedures provided for in paragraph 2 of that Article.

"(4) With regard to Article 9, relating to the transit of personnel, equipment and property to and from the requesting State, the Government of the Kingdom of Saudi Arabia will not be bound to facilitate the transit through its territory of personnel, equipment and property unless there exist diplomatic relations between the Kingdom and the States Parties concerned during the provision of assistance.

"III. Accession to there two Conventions does not in any way imply recognition of Israel and shall not lead to any involvement with Israel on matters governed by these conventions."

(Original in Arabic; translation by the secretariat)

South Africa [10 August 1987]

Convention of Early Notification of a Nuclear Accident

"(a) the Government of the Republic of South Africa does not consider itself bound by either of the dispute settlements provided for in Article 11, paragraph 2, of the Convention,

"(b) the signature of this Convention by the Republic of South Africa in no way implies recognition by South Africa of the United Nations Council for Namibia or its competence to act on behalf of South West Africa/Namibia."

Convention on Assistance in the Case of a Nuclear Accident or Radiological Emergency

"(a) The Government of the Republic of South Africa does not consider itself bound by either of the dispute settlements provided for in Article" 13, paragraph 2, of the Convention,

"(b) The signature of this Convention by the Republic of South Africa in no way implies recognition by South Africa of the United Nations Council for Namibia or its competence to act on behalf of South West Africa/Namibia."

(Original in English)

Spain[11 September 1989]

Convention on Early Notification of a Nuclear Accident

"The Kingdom of Spain does not consider itself bound by the procedures for the settlement of disputes stipulated in paragraph 2 of article 11 of the Convention on Early Notification of a Nuclear Accident."

Convention on Assistance in the Case of a Nuclear Accident or Radiological Emergency,

"The Kingdom of Spain declares that it does not consider itself bound by the provisions in paragraphs 2 and 3 of article 8, in paragraph 2 of article 10, and in paragraph 2 of article 13 of the Convention on Assistance in the Case of a Nuclear Accident of Radiological Emergency."

(Original in Spanish; translation by the Secretariat)

Sri Lanka [11 January 1991]

Convention on Early Notification of a Nuclear Accident

"The Government of the Democratic Socialist Republic of Sri Lanka views Article 1 of the Convention dealing with its scope of application, in the light of the official declarations made by the representatives of China, France, Soviet Union, the United Kingdom and the United States of America as to the readiness of their Governments, on a voluntary basis, to notify the International Atomic Energy Agency and any other States affected by any accident which is not specified in Article 1 of the Convention and which may have transboundary radiological consequences."

Convention on Assistance in the Case of a Nuclear Accident or Radiological Emergency

"(i) The Government of the Democratic Socialist Republic of Sri Lanka considers that the obligations relating to the grant of privileges, immunities and facilities in terms of Article 8, shall be subject to the applicable laws, regulations and procedures of Sri Lanka.

"(ii) The Government of the Democratic Socialist Republic of Sri Lanka declares, in accordance with Article 10.5 that Sri Lanka does not consider itself bound by paragraph 2 of the said Article."

(Original in Singhalese and English)

Sweden [24 June 1992]

Convention on Assistance in the Case of a Nuclear Accident or Radiological Emergency

"Under Article 10 item 5 (b): Sweden reserves the right, notwithstanding what is stated in Article 8 concerning immunities and privileges, to retroactively reclaim reimbursement paid from a tortfeasor acting on behalf of the assisting party, who has caused damage by intent or gross negligence. Sweden will furthermore apply Swedish rules concerning apportionment on grounds of contributory negligence.

"Under Article 8 item 9: Sweden declares that the rules on immunities and privileges in the Convention shall not apply to participants in rescue operations who are Swedish citizens or are resident in Sweden."

(Original in English)

Thailand [21 March 1989]

Convention on Early Notification of a Nuclear Accident

"Thailand does not consider itself bound by both of the dispute settlement procedures provided for in paragraph 2 of Article 11."

Convention on Assistance in the Case of a Nuclear Accident or Radiological Emergency

"Thailand does not consider itself bound by the provisions concerning; (i) the privileges, immunities stipulated in paragraphs 2 and 3 of Article 8; (ii) the claims and compensation prescribed in paragraph 2 of Article 10, and (iii) both of the dispute settlement procedures provided for in paragraph 2 of Article 13."
(Original in English)

Turkey [3 January 1991]

Convention on Early Notification of a Nuclear Accident

"Turkey hereby declares that in accordance with paragraph 3 of the article 11 of the Convention on Early notification of a Nuclear Accident, it does not consider itself bound by the provisions of paragraph 2 of article 11 thereof."

Convention on Assistance in the Case of a Nuclear Accident or Radiological Emergency

"In Conformity with the article 8 paragraph 9 of the Convention on Assistance in the Case of a Nuclear Accident or Radiological Emergency, Turkey does not consider itself bound by article 8, paragraph 2(a) with regard to the immunity from civil proceedings, by paragraph 2(b) concerning exemption from taxation, duties or other charges for personnel of the assisting party,

"Turkey hereby declares that in accordance with article 10, paragraph 5 of the Convention on Assistance in the Case of a Nuclear Accident or Radiological Emergency, it does not consider itself bound by the provisions of paragraph 2 article 10.

"Turkey hereby declares that in accordance with paragraph 3 of the article 13 of the Convention on Assistance in the Case of a Nuclear Accident or Radiological Emergency, it does not consider itself bound by the provisions of paragraph 2 of article 13, thereof." (Original in Turkish; translation provided by the Government)

Ukraine [26 January 1987]

The following identical reservation was received in respect of both Conventions:

"The Ukrainian SSR will not consider itself bound by the provisions of Article II, paragraph 2 of the Convention on Early Notification of a Nuclear Accident and Article 13, paragraph 2 of the Convention on Assistance in the Case of a Nuclear Accident or Radiological Emergency, which envisage the possibility of submitting a dispute between States Parties to arbitration or referring it to the International Court of Justice at the request of any party, and states that for submission of any international dispute to arbitration or referral to the International Court of Justice the agreement of all parties in each individual case is necessary."

(Original in Russian; translation by the Secretariat)

United Arab Emirates [2 October 1987]

Convention on Early Notification of a Nuclear Accident

"The Government of the United Arab Emirates, in accordance with Paragraph 3 of Article 11, does not consider itself bound by the provisions of Paragraph 2 of that article."

Convention on Assistance in the Case of a Nuclear Accident or Radiological Emergency,

"The Government of the United Arab Emirates, in accordance with Paragraph 3 of Article 13, does not consider itself bound by the provisions of Paragraph 2 of that Article."

(Original in English)

United Kingdom [9 February 1990]

Convention on Early Notification of a Nuclear Accident

"The United Kingdom Government affirms that, having regard to Article 3 of the Convention, and as stated by the United Kingdom Secretary of State for Emergency in his address to the Special Session of the General Conference on 24 September 1986, the United Kingdom would in practice notify the IAEA and affected states in the event of an accident to military facilities or equipment which, although not of the type specified in

Article 1 of the Convention, had or might have the consequences specified in that Article"

Convention on Assistance in the Case of a Nuclear Accident or Radiological Emergency

"In pursuance of paragraph 9 of Article 8 of the Convention on Assistance in the Case of a Nuclear Accident or Radiological Emergency, the United Kingdom hereby declares that it considers itself bound by paragraphs 2 and 3 of the said Article 8 to the following extent:

"1. In cases where assistance is provided by the International Atomic Energy Agency, to the extent to which the privileges and immunities provided for in those paragraphs are accorded in the Agreement on the privileges and Immunities of the International Atomic Energy Agency, approved by the Board of Governors on 1 July 1959;

"2. In cases where assistance is provided by any other international intergovernmental organisation, to the extent to which the United Kingdom has agreed to accord the privileges and immunities provided for in those paragraphs;

"3. In cases where assistance is provided by a State Party to the Convention, to the following extent:

"(a) In relation to the State Party providing assistance to the extent that State Party is itself bound by those paragraphs in relation to the United Kingdom;

"(b) the United Kingdom shall only be bound to apply paragraph 2(b) in cases where the State Party is providing assistance without cost to the United Kingdom; and

"(c) the exemption from taxation provided for in paragraph 2(b) shall only extend to an exemption from income tax on the salaries and emoluments of personnel which are paid from the State Party providing assistance and the United Kingdom reserves the right to take those salaries and emoluments into account for the purpose of assessing the amount of taxation to be applied to income from other sources".

(Original in English)

United States of America [19 September 1988]

Convention on Early Notification of a Nuclear Accident

"As provided for in paragraph 3 of article II, the United States declares that it does not consider itself bound by either of the dispute settlement procedures provided for in paragraph 2 of that article."

Convention on Assistance in the Case of a Nuclear Accident or Radiological Emergency

"In accordance with paragraphs 3 and 4 of article 2 of article 7, the United States declares that reimbursement of costs in among the terms of assistance it may provide unless the United States explicitly specifies otherwise or waives reimbursements.

"With respect to any ether State Party that has declared pursuant to paragraph 9 of article 8 that it does not consider itself bound in whole or part by paragraphs 2 or 3, the United States declares pursuant to paragraph 9 that in its treaty relations with that state the United States does not consider itself bound by paragraphs 2 and 3 to the same extent provided in the declaration of that other state party.

"With respect to any other state party that has declared pursuant to paragraph 5 of article 10 that it does not consider itself bound in whole or in part by paragraph 2 or that it will not apply paragraph 2 in whole or in part in cases of gross negligence, the United States declares pursuant to paragraph 5 that in its treaty relations with that state the United States does not consider itself bound by paragraph 2 to the same extent as provided in the declaration of that other state party.

"As provided for in paragraph 3 of article 13, the United States declares that it does not consider itself bound by either of the dispute settlement procedures provided for in paragraph 2 of that article."

(Original in English)

Viet Nam, Socialist Republic Of [29 September 1987]

The following identical reservation was received in respect of both Conventions:

"The Socialist Republic of Viet Nam shall not consider itself to be bound by the provisions of paragraph 2 of article 11 of the Convention on Early Notification of a Nuclear Accident or of paragraph 2 of article 13 of the Convention on Assistance in the Case of a Nuclear Accident or Radiological Emergency, which provide for the possibility of submitting a dispute between States Parties to arbitration or of referring it to the International Court of Justice at the request of any party; the Socialist Republic of Viet Nam declares that the submission of any international dispute to arbitration or its referral to the International Court of Justice requires the agreement of all the parties in each particular case."

(Original in French; translation by the Secretariat)

World Health Organization [10 August 1988]

Convention on Early Notification of a Nuclear Accident

"In conformity with paragraph 5(c) of Article 12, the Director-General of WHO declares that the World Health Organization is competent to act as the directing and co-ordinating authority in international health work in matters covered by the Convention, and to provide related assistance upon the request or acceptance of governments, without prejudice to the national competence of each of its Member States."

Convention on Assistance in the Case of a Nuclear Accident or Radiological Emergency

"In conformity with paragraph 5 (c) of Article 14, the Director-General of WHO declares that the World Health Organization is competent to act as the directing and co-ordinating authority in international health work in matters covered by the Convention, and to provide related assistance upon the request or acceptance of governments, without prejudice to the national competence of each of its member States."

(Original in English)

Work Meteorological Organization [17 April 1990]

Convention on Early Notification of a Nuclear Accident

"I, the undersigned, Prof. G.O.P. Obasi, Secretary-General of the World Meteorological Organization, declare in accordance with paragraph five (c) of Article twelve of the Convention on Early Notification of a Nuclear Accident adopted at Vienna on the twenty-sixth day of September, one thousand nine hundred and eighty six that the World Meteorological Organization has competence in respect of the negotiation, conclusion and application of international agreements in matters covered by this Convention to the extent of fulfilling the purposes of the Organization as laid down in Article two of the Convention of the World Meteorological Organization."

Convention on Assistance in the Case of a Nuclear Accident or Radiological Emergency

"I, the undersigned, Prof. G.O.P. Obasi, Secretary-General

of the World Meteorological Organization, declare in accordance with paragraph five (c) of Article fourteen of the Convention on Assistance in the Case of a Nuclear Accident or Radiological Emergency adopted at Vienna on the twenty-sixth day of September, one thousand nine hundred and eighty-six that the World Meteorological Organization has competence in respect of the negotiation, conclusion and application of international agreements in matters covered by this Convention to the extent of fulfilling the purposes of the Organization as laid down in article two of the Convention of the World Meteorological Organization."

(Original in English)

Reservations/declarations made upon signature

Afghanistan [26 September 1986]

"...the Government of the Democratic Republic of Afghanistan reserves its right to make whatever declaration it deems appropriate at the time of deposit of its instrument of ratification."

(Original in English)

Algeria [24 September 1987]

Article 11. Settlement of Disputes

The People's Democratic Republic Algeria does not consider itself bound by either of the dispute settlement procedures set out in paragraph 2. The People's Democratic Republic of Algeria declares that the submission of any dispute to arbitration or its referral to the International Court of Justice requires the agreement of all parties to the dispute.

Article 12. Entry into force

Algeria's signature will be accompanied by the words 'subject to ratification.'

Article 13. Provisional application

The People's Democratic Republic of Algeria declares that it will apply the Convention provisionally in accordance with Article 13."

Convention on Assistance in the Case of a Nuclear Accident or Radiological Emergency

Article 8. Privileges, immunities and facilities

In accordance with paragraph 9 of Article 8, the people's Democratic Republic of Algeria does not consider itself bound by the provisions of paragraphs 2 and 3 of that Article.

Reservations on paragraph 8

The People's Democratic Republic of Algeria does not consider itself bound by the rules of customary international law.

Article 10. Claims and compensation

The People's Democratic Republic of Algeria states that national law will apply with regard to legal proceedings and compensation.

Article 13. Settlement of disputes

The People's Democratic Republic of Algeria does not consider itself bound by either of the dispute settlement procedures set out in paragraph 2. The People's Democratic Republic of Algeria declares that the submission of any dispute to arbitration or its referral to the International Court of Justice requires the agreement of all parties to the dispute."

Article 14. Entry into force

Algeria's signature will be accompanied by the words 'subject to ratification'.

Article 15. Provisional application

The People's Democratic Republic of Algeria declares

that it will apply the Convention provisionally in accordance with Article 15."

(Original in French; translation by the Secretariat)

Australia [26 September 1986]

"Australia will make any declarations as provided for by the Conventions only upon ratification.

"Attention is also drawn to the statement by the Leader of the Australian delegation to the first Special Session of the General Conference, in particular the sections of the statement which refer to the relationship between the conventions and customary international law."

(Original in English)

Belarus [26 September 1986]

"The Byelorussian SSR also declares that it accepts provisionally the obligations under the conventions in question from the time of their signature and until their ratification. The Byelorussian SSR will not consider itself bound by the provisions of article II, paragraph 2 of the Convention on Early Notification of a Nuclear Accident and article 13, paragraph 2 of the Convention on Assistance in the Case of a Nuclear Accident of Radiological Emergency, which envisage the possibility of submitting a dispute between States Parties to arbitration or referring it to the International Court of Justice at the request of any party, and states that for submission of any international dispute to arbitration or referral to the International Court of Justice the agreement of all parties in each individual case is necessary."

(Original in Russian; translation by the Secretariat)

Bulgaria [26 September 1986]

"From the time of signature and until the conventions come into force for the People's Republic of Bulgaria, the latter will apply both conventions provisionally.

"The People's Republic of Bulgaria does not consider itself bound by the dispute settlement procedures provided for in article II, paragraph 2 of the Convention on Early Notification of a Nuclear Accident and in article 13, paragraph 2 of the Convention on Assistance in the Case of a Nuclear Accident or Radiological Emergency."

(Original in Russian; translation by the Secretariat)

Canada [26 September 1986]

The following identical communication was received in respect of both Conventions :

"...the Government of Canada reserves its right to make whatever declarations it deems appropriate at the time of deposit of its instrument of ratification."

(Original in English)

China [26 September 1986]

Convention on Early Notification of a Nuclear Accident

"1. China does not consider itself bound by either of the dispute settlement procedures provided for in article II, paragraph 2, of the convention.

"2. In view of the urgency of the question of nuclear safety, China accepts article 13, the provisionally applicable clause of the Convention before the Convention's entry into force for China."

Convention on Assistance in the Case of a Nuclear Accident or Radiological Emergency

"1. In cases of gross negligence by the individuals who caused the death, injury, loss or damage, article 10, paragraph 2, of the Convention shall not apply to China.

"2. China does not Consider itself bound by either of the dispute settlement procedures provided for in article 13, paragraph, 2, of the Convention.

"3. In view of the urgency of the question of nuclear

safety, China accepts article 15, the provisionally applicable clause of the Convention before the Convention's Entry into force for China."

(Original in Chinese and English; Supplied by the Government)

Cuba [26 September 1986]

"With regard to the settlement of disputes as described in article II of the Convention on Early Notification of a Nuclear Accident, the Government of Cuba does not consider itself bound by the procedure for referring disputes to the International Court of Justice nor by any decision which the International Court of Justice takes in application of this Convention and which affects Cuba.

"With regard to the settlement of disputes as described in article 13 of the Convention on Assistance in the Case of a Nuclear Accident or Radiological Emergency, the Government of Cuba does not consider itself bound by the procedure for referring dispute to the International Court of Justice nor by any decision which the International Court of Justice takes in application of this Convention and which affects Cuba."

(Original in Spanish; translation by the Secretariat)

Democratic People's Republic Of Korea[29 September 1986]

"1. The Democratic People's Republic of Korea does not consider itself bound by either of dispute settlement procedures provided for in article II, paragraph 2 of the Convention on Early Notification of a Nuclear Accident and in article 13, paragraph 2 of the Convention on Assistance in the Case of a Nuclear Accident or Radiological Emergency.

"2. In view of the urgency of the question of nuclear safety the Democratic People's Republic of Korea will apply both conventions provisionally.

(Original in English)

France [26 September 1986]

Convention on Assistance in the Case of a Nuclear Accident or Radiological Emergency

"Article 8. Privileges, immunities and facilities

The Government of the French Republic declares, in accordance with paragraph 9 of article 8, that France does not consider itself bound by the provisions of paragraphs 2 and 3 of that article.

"Article 10. Claims and compensation

The Government of the French Republic declares, in accordance with paragraph 5 of article 10, that France does not consider itself bound by paragraph 2 of that article.

"Article 13. Settlement of disputes

The Government of the French Republic declares, in accordance with paragraph 3 of article II, that France does not consider, itself bound by the provisions of paragraph 2 of that article."

(Original in French; translation by the Secretariat)

Germany, Federal Of [26 September 1986]

Convention on Early Notification of a Nuclear Accident

"1. With reference to article 13 of the aforementioned Convention, the Federal Republic of Germany will as of today, in accordance with the law applicable in the Federal Republic of Germany, apply the Convention provisionally.

"2. The Federal Republic of Germany is of the view that in the Case of a nuclear accident information about the effects of the accident should also be exchanged between neighbouring States affected by the accident and expresses its wish that also other countries would act accordingly."

Convention on Assistance in the Case of a Nuclear Accident or Radiological Emergency

"...With reference to article 15 of the aforementioned Convention, that the Federal Republic of Germany will as of today, in accordance with the law applicable in the Federal Republic of Germany, apply the Convention provisionally."

(Original in English and German; supplied by the Government)

Greece [September 1986]

The following identical communication was received in respect of both Conventions:

"According to their respective articles 13 and 15, the above two conventions will be provisionally applied in Greece within the framework of the existing internal legislation."

(Original in English)

Hungary [26 September 1986]

Convention on Early Notification of a Nuclear Accident

"The Hungarian People's Republic does not consider itself bound by the dispute settlement procedures provided for in paragraph 2 of article II of the Convention since, in its opinion, the jurisdiction of any arbitral tribunal or of the International Court of Justice can be founded only on the voluntary prior acceptance of such jurisdiction by all the Parties concerned."

Convention on Assistance in the Case of a Nuclear Accident or Radiological Emergency

"The Hungarian People's Republic does not consider itself bound by the dispute settlement procedures provided for in article 13 paragraph 2, of the Convention since, in its opinion, the jurisdiction of any arbitral tribunal or of the International Court of Justice can be founded only on the voluntary prior acceptance of such jurisdiction by all the Parties concerned."

(Original in English and Hungarian; supplied by the Government)

India [29 September 1986]

"While signing the two conventions that were approved by the special session last week, I would like to express the disappointment of my Government that the Convention on Early Notification of a Nuclear Accident does not cover all kinds of accidents. It should have been a full-scope convention covering accidents from whatever source-civil or military, including accidents emanating from nuclear weapons or nuclear weapons tests, since the transboundary effects of radiological safety significance from any source would be equally damaging. Nevertheless, we have decided to sign both conventions, subject to ratification, in view of the solemn assurance that has been given by the five nuclear-weapon States to the effect that they undertake to notify all accidents. This is in keeping with our policy of according to public declarations of State policy equal validity with other international commitments.

"While ratifying these conventions, it is our intention to indicate our reservations with respect to certain articles of the two conventions, as already provided for in them." (Original in English)

Indonesia [26 September 1986]

"The permanent Mission has further the honour to inform the Secretariat that the Government of Indonesia wishes to make the following reservations;

(i) article 13 on Settlement of Disputes of the

Convention on Assistance in the Case of a Nuclear Accident or Radiological Emergency; and

(ii) article II on Settlement of Disputes of the Convention on Early Notification of a Nuclear Accident."

(Original in English)

Iraq [12 August 1987]

Convention on Early Notification of a Nuclear Accident

".....With a reservation on Article 11.2 with regard to the obligation to accept arbitrators to be appointed by the President of the International Court of Justice or the Secretary-General of the United Nations."

Convention on Assistance in the Case of a Nuclear Accident or Radiological Emergency

"1. In accordance with the right under Article 10.5 (a) and (b) of States signing the Convention we make a reservation with regard to Article 8 concerning immunity from legal proceedings, namely that cases of gross negligence shall be excluded from total immunity so that the assisting party shall not be exempted from responsibility;

"2. We make a reservation on Article 13.2 with regard to the obligation to accept arbitrators to be appointed by the President of the International Court of Justice or the Secretary-General of the United Nations."

(Original in Arabic; translation by the Secretariat)

Ireland [26 September, 1986]

"Ireland hereby declares that in accordance with article 8, paragraph 9 of the Convention on Assistance in the Case of a Nuclear Accident or Radiological Emergency, it does not consider itself bound by the provisions of paragraphs 2 & 3 of article 10 thereof."

"Ireland hereby declares that in accordance with article 10, paragraph 5 of the Convention on Assistance in the Case of Nuclear Accident or Radiological Emergency, it does not consider itself found by the provisions of paragraph 2 of article 10 thereof."

(Original in English)

Italy [26 September 1986]

"The Italian Government, upon signature of the Convention on Early Notification of a Nuclear Accident, declares that the clauses of article I are not satisfactory, in so far as they impose on a contracting party the obligation to notify only accidents from which derives the release of radioactive materials which might transcend or has transcended an international boundary, or could have other consequences outside its jurisdiction or control.

"The Italian Government considers that every accident should be notified, also those which have consequences limited to the territory of the State concerned."

(Original in English)

Mongolia [8 January 1987]

The following identical reservation was received in respect of both Conventions:

"The Mongolian People's Republic states that it does not consider itself bound by the provisions of paragraph 2 of article II of the Convention on Early Notification of a Nuclear Accident and the provisions of paragraph 2 of article 13 of the Convention on Assistance in the Case of a Nuclear Accident or Radiological Emergency concerning the procedure of the settlement of disputes arising from the interpretation or application of the Conventions. In its opinion, for submission of any dispute of such nature to arbitration or the International Court of Justice, the Consent of all the parties to the dispute is necessary."

(Original in English and Russian; supplied by the Government)

Netherlands [26 September 1986]

"... declares today on the occasion of the signing of the Convention on Early Notification of a Nuclear Accident, and in accordance with article 13 of that Convention, that his Government, anticipating the entry into force of the Convention for the Kingdom of the Netherlands, with apply its provisions provisionally. This provisional application will come into effect thirty days from today, or, in case the convention will not be in force for at least one other State at that time, on the date on which the Convention will have become applicable to one other State either by means of entry into force or by means of a declaration of provisional application.

"...declares today on the occasion of the signing of the Convention on Assistance in the Case of a Nuclear Accident or Radiological Emergency, and in accordance with article 15 of that Convention, that his Government, anticipating the entry into force of the Convention for the Kingdom of the Netherlands, will apply its provisions provisionally. This provisional application will come into effect thirty days from today, or, in case the Convention will not be in force for at least one other State at that time, on the date on which the Convention will have become applicable to one other State either by means of entry into force or by means of a declaration of provisional application. The provisions of article 10, second paragraph, are being excluded from this provisional application."

(Original in English)

Norway [26 September 1986]

Convention on Assistance in the Case of a Nuclear Accident or Radiological Emergency

"In conformity with article 8, paragraph 9 of the Convention, Norway does not consider herself bound by article 8, paragraph 2(a) as far as immunity from civil proceedings are concerned and by article 8, paragraph 2(b) as far as exemption from taxation, duties or other charges for personnel of the assisting party is concerned.

(Original in English)

Poland [26 September 1986]

"The Government of the Polish People's Republic declares that it will provisionally apply the Convention on Early Notification of a Nuclear Accident adopted at the Special Session of the General Conference of the International Atomic Energy Agency in Vienna on 26 September 1986, with the exception of article II, paragraph 2, over the period between its entry into force and ratification.

"The Government of the Polish People's Republic declares that it will provisionally apply the Convention on Assistance in the Case of a Nuclear Accident or Radiological Emergency adopted in Vienna on 26 September 1986 with the exception of article 13, paragraph 2, over the period between its entry into force and ratification."

(Original in English)

Russian Federation [26 September 1986]

"From the time of signature and until the conventions come into force for the USSR, the latter will apply both conventions provisionally.

"The USSR will not consider itself bound by the provisions of article II, paragraph 2 of the convention on Early Notification of a Nuclear Accident and article 13, paragraph 2 of the Convention on Assistance in the Case of a Nuclear Accident or Radiological Emergency, which envisage the possibility of submitting a dispute between States parties to arbitration or referring it to the International Court of Justice at the request of any party, and states that for the submission of any international dispute to arbitration or referral to the International Court of Justice the agreement of all parties in each individual case is necessary."

(Original in Russian; translation by the Secretariat)

Thailand [25 September 1987]

Convention on Early Notification of a Nuclear Accident

"In accordance with paragraph 3 of article 11 of the Convention on Early Notification of a Nuclear Accident, Thailand does not consider itself bound by the dispute settlement procedures provided for in paragraph 2 of that Article."

Convention on Assistance in the Case of a Nuclear Accident or Radiological Emergency

"In accordance with paragraph 9 of Article 8 and paragraph 5 of Article 10 of the Convention on Assistance in the Case of a Nuclear Accident or Radiological Emergency, Thailand does not consider itself bound by the provisions stipulated in paragraphs 2 and 3 of Article 8 and paragraph 2 of Article 10, and also declares that, in accordance with paragraph 3 of Article 13 of the Convention, it does not consider itself bound by the dispute settlement procedures provided for in paragraph 2 of that Article."

(Original in English)

Turkey [26 September 1986]

"......declarations or reservations will be made, if any, on article II of the Convention on Early Notification of a Nuclear Accident and on articles 8, 9 and 13 of the

Convention on Assistance in the Case of a Nuclear Accident or Radiological Emergency during the course of the submission of the instrument of ratification to the depositary."

(Original in English)

Ukraine [26 September 1986]

"The Ukrainian SSR also declares that it accepts provisionally the obligations under the Conventions in question from the time of their signature and until there ratification. The Ukrainian SSR will not consider itself bound by the provisions of article II, paragraph 2 of the Convention on Early Notification of Nuclear Accident and article 13, paragraph 2 of the Convention on Assistance in the Case of a Nuclear Accident or Radiological Emergency, which envisage the possibility of submitting a dispute between States Parties to arbitration or referring it to the International Court of Justice at the request of any party, and states that for the submission of any international dispute to arbitration or referral to the International Court of Justice the agreement of all parties in each individual case is necessary."

(Original in Russian; translation by the Secretariat)

United Kingdom of Great Britain And Northern Ireland
[26 September 1986]

Convention on Early Notification of a Nuclear Accident

"The, United Kingdom will apply this Convention provisionally from today's date to the extent permitted by its existing laws, regulations and administrative arrangements.

"The United Kingdom Resident Representative affirms that, having regard to article 3 of the Convention, and as stated by the United Kingdom Secretary of State for Energy in his address to the Special Session of the

General Conference on 24 September, the United Kingdom would in practice notify the IAEA and affected States in the event of an accident to military facilities or equipment which, although not of the type specified in article I of the Convention, had or might have the consequence specified in that article."

Convention on Assistance in the Case of a Nuclear Accident or Radiological Emergency

"The United Kingdom will apply this Convention provisionally form today's date to the extent permitted by its existing laws, regulations and administrative arrangements."

(Original in English)

United States of America [26 September 1986]

Convention on Early Notification of a Nuclear Accident

"As provided for in paragraphs 3 of article II, the United States declares that it does not consider itself bound by either of the dispute settlement procedures provided for in paragraph 2 of that article."

Convention on Assistance in the Case of a Nuclear Accident or Radiological Emergency

"In accordance with paragraphs 3 and 4 of article 2 and paragraph 2 of article 7, the United States declares that reimbursement of costs is among the terms of assistance it may provide unless the United States explicitly specifies otherwise or waives reimbursement.

"With respect to any other State Party that has declared pursuant to paragraph 9 of article 8 that it does not consider itself bound in whole or in part by paragraphs 2 or 3, the United States declares pursuant to paragraph 9 that in its treaty relations with that State the United States does not consider itself bound by paragraphs 2 and 3 to the same extent provided in the declaration of that other State Party.

"With respect to any other State Party that has declared pursuant to paragraph 5 of article 10 that it does not consider itself bound in whole or in part by paragraph 2 or that it will not apply paragraph 2 in whole or in part in cases of gross negligence, the United States does not consider itself bound by paragraph 2 to the same extent as provided in the declaration of that other State party.

"As provided for in paragraph 3 of article 13, the United States declares that it does not consider itself bound by either of the dispute settlement procedures provided for in paragraph 2 of that article."

(Original in English)

6 Treaty Banning Nuclear Weapon Tests in the Atmosphere, in Outer Space and Under Water (1963)

The Governments of the United States of America, the United Kingdom of Great Britain and Northern Ireland, and the Union of Soviet Socialist Republics, hereinafter referred to as the "Original Parties",

Proclaiming as their principal aim the speediest possible achievement of an agreement on general and complete disarmament under strict international control in accordance with the objectives of the United Nations which would put an end to the armaments race and eliminate the Incentive to the production and testing of all kinds of weapons, including nuclear weapons,

Seeking to achieve the discontinuance of all test explosions of nuclear weapons for all time, determined to continue negotiations to this end, and desiring to put an end to the contamination of man's environment by radioactive substances,

Have agreed as follows:

Article I

1. Each of the Parties to this Treaty undertakes to prohibit, to prevent, and not to carry out any nuclear weapon test explosion, or any other nuclear

explosion, at any place under its jurisdiction or control:

(a) in the atmosphere; beyond its limits, including outer space; or under water, including territorial waters or high seas; or

(b) in any other environment if such explosion causes radioactive debris to be present outside the territorial limits of the States under whose jurisdiction or control such explosion is conducted. It is understood in this connection that the provisions of this sub-paragraph are without prejudice to the conclusion of a treaty resulting in the permanent banning of all nuclear test explosions, including all suçh explosions underground, the conclusion of which, as the Parties have stated in the Preamble to this Treaty, they seek to achieve.

2. Each of the parties to this Treaty undertakes furthermore to refrain from causing, encouraging, or in any way participating in, the carrying out of any nuclear weapon test explosion, or any other nuclear explosion, anywhere which would take place in any of the environments described, or have the effect referred to, in paragraph 1 of this Article.

Article II

1. Any Party may propose amendments to this Treaty. The test of any proposed amendment shall be submitted to the Depositary Governments will shall circulate it to all Parties to this Treaty. Thereafter, if requested to do so by one-third or more of the Parties, the Depositary Governments shall convene a conference, to which they shall invite all of the Parties, to consider such amendment.

2. Any amendment to this Treaty must be approved by a majority of the votes of all the Parties to this

Treaty, including the votes of all of the Original Parties. The amendment shall enter into force for all Parties upon the deposit of instruments of ratification by a majority of all the Parties, including the instruments of ratification of all the Original Parties.

Article III

1. This Treaty shall be open to all States for signature. Any State which does not sign this Treaty before its entry into force in accordance with paragraph 3 of this Article may accede to it at any time.
2. This Treaty shall be subject to ratification by signatory States. Instruments of ratification and instruments of accession shall be deposited with the Governments of the Original Parties—the United States of America, the United Kingdom of Great Britain and Northern Ireland, and the Union of Soviet Socialist Republics—which are hereby designated the Depositary Governments.
3. This Treaty shall enter into force after its ratification by all the Original Parties and the deposit of their instruments of ratification.
4. For States whose instruments of ratification or accession are deposited subsequent to the entry into force of this Treaty, it shall enter into force on the date of the deposit of their instruments of ratification or accession.
5. The Depositary Governments shall promptly inform all signatory and acceding States of the date of each signature, the date of deposit of each instrument of ratification of and accession of Treaty, the date of its entry into force, and the date of receipt of any requests for conferences or other notices.
6. This Treaty shall be registered by the Depositary Governments pursuant to Article 102 of the Charter of the United Nations.

Article IV

This Treaty shall be of unlimited duration

Each Party shall in exercising its national sovereignty have the right to withdraw from the Treaty if it decides that extraordinary events, related to the subject matter of this Treaty, have jeopardized the supreme interests of its country. It shall give notice of such withdrawal to all other Parties to the Treaty three months in advance.

Article V

This Treaty, of which the English and Russian texts are equally authentic, shall be deposited in the archives of the Depositary Governments Duly certified copies of this Treaty shall be transmitted by the Depositary Governments to the Governments of the signatory and acceding States.

In witness where of the undersigned, duly authorized, have signed this Treaty

Done in triplicate at the city of Moscow the fifth day of August, one thousand nine hundred and sixty-three.

7 Convention on the Physical Protection of Nuclear Material (1993)

Reservations/Declarations made upon or following expressing consent to be bound:

Argentina [6 April 1989]

"In accordance with the provisions of Article 17.3 of the Convention, Argentina does not consider itself bound by either of the dispute settlement procedures provided for in Article 17.2 of the Convention."

(Original in Spanish, translation by the Secretariat)

Bulgaria [10 April 1984]

"The People's, Republic of Bulgaria does not consider itself bound by Article 17(2) of the Convention on the Physical Protection of Nuclear Material, according to which any dispute concerning the interpretation or application of the Convention shall, at the request of any party to such dispute, be submitted to arbitration or referred to the International Court of Justice."

(Originals in English and Bulgarian; supplied by the Government)

China [10 January 1989]

"China will not be bound by the two dispute

settlement procedures as stipulated in Paragraph 2, Article 17 of the Said Convention."

(Original in Chinese; translation by the Secretariat)

European Atomic Energy Community

[6 September 1991]

"Pursuant to Article 18 (4) (c) of the Convention, [the European Atomic Energy Community] would like to declare:

(a) that the Member States of the Community are at present Belgium, Denmark, France, Germany, Greece, Ireland, Italy, Luxembourg, the Netherlands, Portugal, Spain, and the United Kingdom of Great Britain and Northern Ireland;

(b) that Articles 7 to 13 of the Convention are not applicable to the Community.

"Further, pursuant to Article 17 (3) of the Convention, [the European Atomic Energy Community] declare[s] that, since only States may be parties in cases before the International Court of Justice, the Community considers itself exclusively bound by the arbitration procedures provided for in Article, 17 (2)."

(Original in English)

France [6 September 1991]

"(1) In approving the Convention, the French Government expresses the following reservation: the offences described in sub-paragraphs 1(e) and 1(f) of Article 7 of the Convention shall be punished in accordance with the provisions of French penal legislation.

"(2) The French Government declares that the jurisdiction referred to in Article 8, paragraph 4 may not be invoked against it, since the criterion of jurisdiction based on involvement in international nuclear transport as the exporting or importing State is not expressly recognized in international law and is not provided for in French national legislation.

"(3) In accordance with Article 17, paragraph 3, France declares that it does not accept the competence of the International Court of Justice in the settlement of the disputes referred to in paragraph 2 of this article, nor that of the President of the International Court of Justice to appoint one or more arbitrators."

(Original in French; Translation by the Secretariat)

Guatemala [23 April 1985]

"The Republic of Guatemala does not consider itself bound by any of the dispute settlement procedures set out in paragraph 2 of Article 17 of the Convention 'which provide for the submission of disputes to arbitration or their referral to the International Court of Justice for decision."

(Original in Spanish; translation by the Secretariat)

Hungary [4 May 1984]

(The following reservation was withdrawn 30 Nov. 1989)

Confirms the reservation made upon signature.

(Originals in English and Hungarian; supplied by the Government)

Indonesia [5 November 1986]

"The Government of the Republic of Indonesia does not consider itself bound by .the provision of Article 17, paragraph 2 of this Convention and take the position that any dispute relating to the interpretation or application of the Convention may only be submitted to arbitration or to the International Court of Justice with the agreement of all the parties to the dispute."

(Originals in English and Indonesian; supplied by the Government)

Italy [6 September 1991]

Confirms the reservations and declaration made upon signature.

(Original in English)

Korea, Republic of [7 April 1982]

Confirms the reservation made upon signature.

(Original in English)

Mongolia [28 May 1986]

(The following reservation was withdrawn 18 June 1990)
Confirms the reservation made upon signature.

(Originals in English and Mongolian; supplied by the Government)

Netherlands [6 September 1991]

"With regard to the obligation to exercise jurisdiction referred to in Article 10 of the Convention on the Physical Protection of Nuclear Material, done at Vienna/New York on 3 March 1980, the Kingdom of the Netherlands makes the reservation, that in cases where the judicial authorities of the Netherlands are unable to exercise jurisdiction on the grounds of one of the principles referred to in Article 8, paragraph 1, of the Convention, the Kingdom shall be bound by this obligation only if it has received an extradition request from a Party to the convention and the said request has been rejected."

(Original in English)

Poland [5 October 1983]

Confirms the reservation made upon signature.

(Original in French; translation by the Secretariat)

Russian Federation [25 May 1983]

Confirms the reservation made upon signature.

(Original in Russian; translation by the Secretariat)

Spain [6 September 1991]

"The Kingdom of Spain declares, in accordance with paragraph 3 of Article 17 of the Convention, that it does not consider itself bound by the procedure for the settlement of disputes stipulated in paragraph 2 of Article 17."

(Original in Spanish; translation by the Secretariat)

Turkey [27 February 1985]

Confirms the reservation made upon signature.

(Original in English)

United Kingdom [11 December 1991]

"...the Convention was extended to cover the Bailiwicks of Jersey and Guernsey and the Isle of man with effect from 6 October 1991. The United Kingdom's Instrument of Ratification should accordingly be construed to extend to them."

(Original in English)

Reservations/declarations made upon signature

Argentina [28 February 1986]

"In accordance with the provision of Article 17.3, the Republic of Argentina does not consider itself bound by any of the arbitration procedures laid down in Article 17.2 of the Convention."

(Original in Spanish; translation by the Secretariat)

Bulgaria [23 June 1981]

"... with the reservation that the people's Republic of Bulgaria does not consider itself bound by Article 17, paragraph 2, of the said Convention."

(Originals in French and Bulgarian; translation by the Secretariat)

European Atomic Energy Community [23 June 1980]

"At Present the following States are members of the European Atomic Energy Community: Belgium, Denmark, France, the Federal Republic of Germany, Ireland, Italy, Luxembourg, the Netherlands and the United Kingdom.

"In signing the convention, the community declares that, when it has deposited the instrument of approval or acceptance pursuant to Article 18 and the Convention has entered into force for the Community, Pursuant to Article 19, Articles 7 to 13 of the Convention will not apply to it.

"Furthermore, the Community declares that, because under Article 34 of the Statute of the International Court of Justice only States may be parties in cases before the Court, it can only be bound by the arbitration procedure set out in Article 17(2)."

(Original in English)

France [13 June 1980]

"Recalling its statement contained in document CPNM/90 of 25 October 1979, the French Government declares that the jurisdiction referred to in Article 8, paragraph 4 may not be invoked against it, since the criterion of jurisdiction based on involvement in international nuclear transport as the exporting or importing State is not expressly recognized in international law and is not provided for in French national legislation."

"In accordance with Article 17, paragraph 3, France declares that it does not accept the competence of the International Court of Justice in/the settlement of the disputes referred to in paragraph 2 of this article, nor that of the President of the International Court of Justice to a point one or more arbitrators."

(Original in French; Translation by the Secretariat)

Hungary [17 June 1980]

"The Hungarian People's Republic does not consider itself bound by Article 17, paragraph 2, of the Convention, Which provides that any dispute of this character which cannot be settled in the manner prescribed in paragraph 1 shall, at the request of any party to such dispute, be submitted to arbitration or referred to the International Court of Justice for decision."

(Originals in English and Hungarian; supplied by the Government)

Israel [17 June 1983]

"In accordance with Article 17, paragraph 3, Israel declares that it does not consider itself bound by the dispute settlement procedures provided for in paragraph 2 of Article 17."

(Original in English)

Italy [13 June 1980]

"1) In connection with Art. 4.2

Italy considers that if assurances as to the levels of physical protection described in annex I have not been received in good time the importing state party may take appropriate bilateral steps as far as practicable to assure itself that the transport will take place in compliance with the aforesaid levels.

"2) In connection with Art. 10

The last words' through proceedings in accordance with the laws of the State are to be considered as referring to the whole Article 10.

"Italy considers that international co-operation and assistance for physical protection and recovery of nuclear materials as well as criminal rules and extradition will apply also to the domestic use, storage and transport of nuclear material used for peaceful purposes. Italy also considers that no provision contained in this convention shall be interpreted as precluding the possibility to widen the scope of the convention at the review conference foreseen in Art. 16."

(Original in English)

Korea, Republic of [29 December 1986]

"...the Government of the Republic of Korea does not consider itself bound by the dispute settlement procedures provided for in paragraph 2 of Article, 17."

(Original in English)

Mongolia [23 January 1986]

"...does not consider itself bound by the provisions of para. 2 Article 17 of the Convention, whereby disputes arising out of the interpretation or application of the Convention could be submitted to arbitration or referred to the International Court of Justice for decision at the request of any party to the dispute."

(Original in English)

Poland [6 August 1980]

"The Polish People's Republic does not consider itself bound by the procedure for settlement of disputes laid down in Article 17.2 of the Convention."

(Originals in French and Polish; translation by the Secretariat)

Romania [15 January 1981]

"The Socialist Republic of Romania declares that it does not consider itself bound by the provisions of Article 17, paragraph 2 of the Convention on the Physical Protection of Nuclear Material, which state that any dispute concerning the interpretation or application of the Convention which cannot be settled by negotiation or by any other peaceful means of settling disputes shall, at the request of any party to such dispute, be submitted to arbitration or referred to the International Court of Justice for decision.

"The Socialist Republic of Romania considers that such disputes can be submitted to arbitration or to the International Court of Justice only with the consent of all parties to the dispute in each individual case.

"In signing the Convention on the Physical Protection of Nuclear Material, the Socialist Republic of Romania declares that, in its interpretation, the provisions of Article 18, paragraph 4 refer exclusively to organizations to which the Member States have transferred competence to negotiate, conclude and apply international agreements on their behalf and to exercise the rights and fulfil the responsibilities entailed by such agreements including the right to vote."

(Original in French; translation by the Secretariat)

Russian Federation [22 May 1980]

"The Union of Soviet Socialist Republics does not consider itself bound by the provisions of Article 17, paragraph 2 of the Convention that any dispute concerning the interpretation or application of the Convention shall be submitted to arbitration or referred to the International Court of Justice at the request of any party to such dispute."

(Original in Russian; translation by the Secretariat)

South Africa [18 May 1981]

"In accordance with Article 17, paragraph 3, the Republic of South Africa declares that it does not consider itself bound by the dispute settlement procedures provided for in paragraph 2 of Article 17."

(Original in English)

Spain[7 April 1986]

"... in accordance with paragraph 3 of Article 17 of the Convention, Spain does not consider itself bound by the procedure for the settlement of disputes stipulated in paragraph 2 of Article 17."

(Original in Spanish; translation by the Secretariat)

Turkey [23 August 1983]

"Turkey, in accordance with Article 17, paragraph 3, of the Convention does not consider itself bound by Article 17, Paragraph 2 of the Convention."

(Original in English)

8 Treaty on the Non-Proliferation of Nuclear Weapons

The States concluding this Treaty, hereinafter referred to as the "Parties to the Treaty".

Considering the devastation that would be visited upon all mankind by a nuclear war and the consequent need to make effort to avert the danger of such a war and to take measures to safeguard the security of peoples.

Believing that the proliferation of nuclear weapons would seriously enhance the danger of nuclear war.

In conformity with resolutions of the United Nations General Assembly calling for the conclusion of an agreement on the prevention of wider dissemination of nuclear weapons,

Undertaking to cooperate in facilitating the application of International Atomic Energy Agency safeguards on peaceful nuclear activities.

Expressing their support for research, development and other efforts to further the application, within the framework of the International Atomic Energy Agency safeguard system, of the principle of safeguarding effectively the flow of source and special fissionable materials by use of instruments and other techniques at certain strategic points.

Affirming the principle that the benefits of peaceful applications of nuclear technology, including any technological by-products which may be derived by nuclear-weapon States from the development of nuclear explosive devices, should be available for peaceful purposes to all Parties to the Treaty, whether nuclear-weapon or non-weapon States,

Convinced that, in furtherance of this principle, all Parties to the Treaty are entitled to participate in the fullest possible exchange of scientific information for, and to contribute alone or in cooperation wit h other States to, the further development of the applications of atomic energy for peaceful purposes,

Declaring their intention to achieve at the earliest possible date the cessation of the nuclear arms race and to undertake effective measures in the direction of nuclear disarmament.

Urging the cooperation of all States in the attainment of this objective.

Recalling the determination expressed by the Parties to the 1963 Treaty banning nuclear weapon tests in the atmosphere in outer space and under water in its Preamble to seek to achieve the discontinuance of all test explosions of nuclear weapons for all time and to continue negotiations to this end.

Desiring to further the easing of international tension and the strengthening of trust between States in order to facilitate the cessation of the manufacture of nuclear weapons, the liquidation of all their existing stockpiles, and the elimination from national arsenals of nuclear weapons and the means of their delivery pursuant to a treaty on general and complete disarmament under strict and effective international control.

Recalling that, in accordance with the Charter of the United Nations, States must refrain in their international relations from the threat or use of force against the territorial integrity or political independence of any State, or in any other manner inconsistent with the Purposes of the United Nations, and that the establishment and maintenance of international peace and security are to be promoted with the diversion for armaments of the world's human and economic resources.

Have agreed as follows:

Article I

Each nuclear-weapon State Party to the Treaty undertakes not to transfer to any recipient whatsoever nuclear weapons or other nuclear explosive devices or control over such weapons or explosive devices directly or indirectly; and not in any way to assist, encourage, or induce any non-nuclear-weapon State to manufacture or otherwise acquire nuclear weapons or other nuclear explosive devices, or control over such weapons or explosive devices.

Article II

Each non-nuclear-weapon State Party to the Treaty undertakes not to receive the transfer from any transferrer whatsoever of nuclear weapons or other nuclear explosive devices or of control over such weapons or explosive devices directly, or indirectly; not to manufacture or otherwise acquire nuclear weapons or other nuclear explosive devices; and to seek or receive any assistance in the manufacture of nuclear weapons or other nuclear explosive devices.

Article III

1. Each non-nuclear-weapon State Party to the Treaty undertakes to accept safeguards, as set forth in an agreement to be negotiated and concluded with the International Atomic Energy Agency in accordance with the Statute of the International Atomic Energy Agency and the Agency's safeguards system, for the exclusive purpose of verification of the fulfilment of

its obligations assumed under this Treaty with a view to preventing diversion of nuclear energy from peaceful uses to nuclear weapons or other nuclear explosive devices. Procedures for the safeguards required by this article shall be followed with respect to source or special fissionable material whether it is being produced, processed or used in any principal nuclear facility or is outside any such facility of such State, under its jurisdiction, or carried out under its control anywhere.

2. Each State Party to the Treaty undertakes not to provide; (a) source or special fissionable material, or (b) equipment or material especially designed or prepared for the processing, use or production of special fissionable material, to any non-nuclear-weapon State for peaceful purposes unless the source or special fissionable material shall be subject to the safeguards required by this article.

3. The safeguards required by this article shall be implemented in a manner designed to comply with article IV of this Treaty, and to avoid hampering the economic or technological development of the Parties or international cooperation in the field of peaceful nuclear activities, including the international exchange of nuclear material and equipment for the processing, use or production of nuclear material for peaceful purposes in accordance with the provisions of this article and the principle of safeguarding set forth in the Preamble of the Treaty.

4. Non-nuclear-weapon States Party to the Treaty shall conclude agreements with the International Atomic Energy Agency to meet the requirements of this article either individually or together with other States in accordance with the Statute of the International Atomic Energy Agency. Negotiation of

such agreements shall commence within 180 days from the original entry into force of this Treaty. For States depositing their instruments of ratification or accession after the 180-day period, negotiation of such enter into force not later than eighteen months after the initiation of negotiations.

Article IV

1. Nothing in this Treaty shall be interpreted as affecting the inalienable right of all Parties to the Treaty to develop research, production and use of nuclear energy for peaceful purposes without discrimination and in conformity with articles I and II of this Treaty.
2. All the Parties to the Treaty undertake to facilitate, and have the right to participate in, the fullest possible exchange of equipment, materials and scientific and technological information for the peaceful uses of nuclear energy. Parties to the Treaty in a position to do so shall also cooperate in contributing alone or together with other States or international organizations to the further development of the applications of nuclear energy for peaceful purposes, especially in the territories for the needs of the developing areas of the world.

Article V

Each Party to the Treaty undertakes to take appropriate measures to ensure that, in accordance with this Treaty, under appropriate international observation and through appropriate international procedures, potential benefits from any peaceful applications of nuclear explosions will be made available to non-nuclear-weapon States Party to the Treaty on a non-discriminatory basic and that the charge to such Parties for the explosive devices used will be as low as possible and exclude any charge for research and development. Non-nuclear-weapon States Party to the Treaty shall be able to obtain such benefits, pursuant to a special international agreement or agreements, through an appropriate international body with adequate representation of non-nuclear-weapon States. Negotiations on this subject shall commence as soon as possible after the Treaty enters into force. Non-nuclear-weapon States Party to the Treaty so desiring may also obtain such benefits pursuant to bilateral agreements.

Article VI

Each of the Parties to the Treaty undertakes to pursue negotiations in good faith on effective measures relating to cessation of the nuclear arms race at an early date and to nuclear disarmament, an on a treaty on general and complete disarmament under strict and effective international control.

Article VII

Nothing in this Treaty affects the rights of any group of States to conclude regional treaties in order to assure the total absence of nuclear weapons in their respective territories.

Article VIII

1. Any Party to the Treaty may propose amendments to this Treaty. The text of any proposed amendment shall be submitted to the Depositary Governments which shall circulate it to all Parties to the Treaty. Thereupon, if requested to do so by one-third or more of the Parties to the Treaty, the Depositary Governments shall convene a conference, to which they shall invite all the Parties to the Treaty, to consider such an amendment.
2. Any amendment to this Treaty must be approved by a majority of the votes of all the Parties to the Treaty, including the votes of all nuclear-weapon

States party to the Treaty and all other Parties which, on the date the amendment is circulated, are members of the Board of Governors of the International Atomic Energy Agency. The amendment shall enter into force for each Party that deposit its instrument of ratification of the amendment upon the deposit of such instruments of ratification of all nuclear-weapon States Party to the Treaty and all other Parties which, on the date the amendment is circulated, are members of the Board of Governors of the International Atomic Energy Agency. Thereafter, it shall enter into force for any other Party upon the deposit of its instrument of ratification of the amendment.

3. Five years after the entry into force of this Treaty, a conference of Parties to the Treaty shall be held in Geneva, Switzerland, in order to review the operation of this Treaty with a view to assuring that the purposes of the Preamble and the provisions of the Treaty are being realized. At intervals of five years thereafter, a majority of the Parties to the Treaty may obtain, by submitting a proposal to this effect to the Depositary Governments, the convening of further conferences with the same objective of reviewing the operation of the Treaty.

Article IX

1. This Treaty shall be open to all State for signature. Any State which does not sign the Treaty before its entry into force in accordance with paragraph 3 of this article may accede to it at any time.
2. This Treaty shall be subject to ratification by signatory States. Instruments of ratification and instruments of accession shall be deposited with the Governments of the United States of America, the United Kingdom of Great Britain and Northern

Ireland and the Union of Soviet Socialist Republics, which are hereby designated the Depositary Governments.

3. This Treaty shall enter into force after its ratification by the States, the Governments of which are designated Depositaries of the Treaty, and forty other States signatory to this Treaty and the deposit of their instruments of ratification. For the purposes of this Treaty, a nuclear-weapon State is one which has manufactured and exploded a nuclear weapon or other nuclear explosive devise prior to January 1, 1967.
4. For States whose instruments of ratification or accession are deposited subsequent to the entry into force of this Treaty. It shall enter into force on the date of the deposit of their instruments of ratification or accession.
5. The Depositary Governments shall promptly inform all signatory and acceding States of the date of each signature, the date of deposit of each instrument of ratification or of accession, the date of the entry into force of this Treaty, and the date of receipt of any requests for convening a conference or other notices.
6. This Treaty shall be registered by the Depository pursuant to article 102 of the Charter of the United Nations.

Article X

1. Each Party shall in exercising its national sovereignty have the right to withdraw from the Treaty if it decides that extraordinary events, related to the subject matter of this Treaty, have jeopardized the supreme interests of its country. It shall give notice of such withdrawal to all other Parties to the Treaty and to the United Nations Security Council three months in advance. Such notice shall include a

settlement of the extraordinary events it regards as having jeopardized its supreme interests.

2. Twenty-five years after the entry into force of the Treaty, a conference shall be convened to decide whether the Treaty shall continue in force indefinitely, or shall be extended for an additional fixed period or periods. This decision shall be taken by a majority of the Parties to the Treaty.

Article XI

This Treaty, the English, Russian, French, Spanish and Chinese texts of which are equally authentic, shall be deposited in the archives of the depositary Governments. Duly certified copies of this Treaty shall be transmitted by the Depositary Governments to the Governments of the signatory and acceding States.

9 Additional Protocol II to the Treaty for the Prohibition of Nuclear Weapons in Latin America (1968)

By the President of the United States of America

A Proclamation

Considering that:

Additional Protocol II to the Treaty for the Prohibition of Nuclear Weapons in Latin America, done at the City of Mexico on February 14, 1967, was signed on behalf of the United States of America on April 1, 1968, the text of which Protocol is word for word as follows:

Additional Protocol II

The undersigned Plenipotentiaries, furnished with full powers by their respective Governments,

Convinced that the Treaty for the Prohibition of Nuclear Weapons in Latin America, negotiated and signed in accordance with the recommendations of the General Assembly of the United Nations in Resolution 1911 (XVIII) of 27 November 1963, represents an important step towards ensuring the non-proliferation of nuclear weapons,

Aware that the non-proliferation of nuclear weapons is not an end in itself but, rather, a means of achieving general and complete disarmament at a later stage, and

Desiring to contribute, so far as lies in their power, towards ending the armaments race, especially in the field of nuclear weapons, and towards promoting and strengthening a world at peace, based on mutual respect and sovereign equality of States,

Have agreed as follows:

Article 1.

The statute of denuclearization of Latin America in respect of warlike purposes, as defined, delimited and set forth in the Treaty for the Prohibition of Nuclear Weapons in Latin America of which this instrument is an annex, shall be fully respected by the Parties to this Protocol in all its express aims and provisions.

Article 2.

The Governments represented by the undersigned Plenipotentiaries undertake, therefore, not to contribute in anyway to the performance acts involving a violation of the obligations of article 1 of the Treaty in the territories to which the Treaty applies in accordance with article 4 thereof.

Article 3.

The Governments represented by the undersigned Plenipotentiaries to undertake not to use or threaten to use nuclear weapons against the Connecting Parties of the Treaty for the Prohibition of Nuclear Weapons in Latin America.

Article 4.

The duration of this Protocol shall be the same as that of the treaty for the Prohibition of Nuclear Weapons in Latin America of which this Protocol is an annex, and the definitions of territory and nuclear weapons set the in articles 3 and 5 of the Treaty shall be applicable to this Protocol, as well as the provisions regarding ratification, reservations, denunciation, authentic its and registration contained in articles 26, 27, 30 and 31 of the Treaty.

Article 5.

This Protocol shall enter into force, for the States which have defied it, on the date of the deposit of their respective instruments of ratification.

In Witness Whereof the undersigned Plenipotentiaries, having deposited or full powers, found in good and due form, sign this Additional Protocol on behalf of their respective Governments.

The Senate of the United States of America by its resolution of April 19, 1971, one thirds of the Senators present concurring, gave its advice and consent to the ratification of Additional Protocol II, with the following understandings and declarations:

I

That the United States Government understands the reference in Article 3 the treaty to "its own legislation: to relate only to such legislation as is compatible with the rules of international law and as involves an exercise of sovereignty consistent with those rules, and accordingly that ratification of Additional Protocol II by the United States Government could not be regarded complying recognition, for the purposes of this treaty and its protocols or for other purpose, of any legislation which did not, in the view of the United States, comply with the relevant rules of international law.

That the United States Government takes note of the Preparatory Commissions interpretation of the treaty, as set forth in the Final Act, that, governed by principles and rules of international law, each of the Contracting Parties has exclusive power and legal competence, unaffected by the terms of the treaty, to grant or deny non-Contracting Parties transit and transport privileges.

That as regards the undertaking in Article 3 of Protocol II not to use or often to use nuclear weapons against the Contracting Parties, the United States Government would have to consider that an armed attack by a contracting Party, in which it was assisted by a nuclear-weapon state, would be incomplete with the Contracting Party's corresponding obligations under Article I of treaty.

II

That the United States Government considers that the technology of making war explosive devices for peaceful purposes is indistinguishable from the technology of making nuclear weapons, and that nuclear weapons and nuclear explosive devices for peaceful purposes are both capable of releasing nuclear energy uncontrolled manner and have the common group of characteristics of large amounts of energy generated instantaneously from a compact source. Therefore, the United States Government understands the definition contained in Article 5 of the treaty as necessarily encompassing all nuclear explosive devices. It is also understood that Articles 1 and 5 restrict accordingly the activities of the Contracting Parties under paragraph 1 of Article 18.

That the United States Government understands that paragraph 4 of Article 18 of the treaty permits, and that United States adherence to Protocol II will not prevent, collaboration by the United States with Contracting Parties for the purpose of carrying out explosions of nuclear devices for peaceful purposes in a manner consistent with a policy of not contributing to the proliferation of nuclear weapons capabilities. In this connection, the United States Government notes Article V of the Treaty on the Non-Proliferation of Nuclear Weapons, under which it joined in an undertaking to take appropriate measures to ensure that potential benefits of peaceful applications of nuclear explosions would be made available to non-nuclear-weapon states party to that treaty, and reaffirms its willingness to extend such undertaking, on the same basis, to states precluded by the present treaty from manufacturing or acquiring any nuclear explosive device.

III

That the United States Government also declares that, although not required by Protocol II, it will act with respect to such territories of Protocol I adherents as are with in the geographical area defined in paragraph 2 of Article 4 of the treaty in the same manner as Protocol II requires it to act with respect to the territories of Contracting Parties.

The President ratified Additional Protocol II on May 8, 1971, with the above-recited understandings and declarations, in pursuance of the advice and consent of the Senate.

It is provided in Article 5 of Additional Protocol II that the Protocol shall enter into force, for the States which have ratified it, on the date of the deposit of their respective instruments of ratification.

The instrument of ratification of the United Kingdom of Great Britain and Northern Ireland was deposited on December 11, 1969 with understandings and a declaration, and the instrument of ratification of the United States of America was deposited on May 12, 1971 with the above-recited understandings and declarations.

In accordance with Article 5 of the Additional Protocol II, the Protocol entered into force for the United States of America on May 12, 1971, subject to the above recited understandings and declarations.

Now, therefore I, Richard Nixon, President of the United States of America, proclaim and make public Additional Protocol II to the Treaty for the Prohibition of Nuclear Weapons in Latin America to the end that it shall be observed and fulfilled with good faith, subject to the above-recited understandings and declarations, on and after May 12, 1971 by the United States of America and by the citizens of the United States of America and all other persons subject to the jurisdiction thereof.

In testimony whereof. I have signed this proclamation and caused the Seal of the United States of America to be affixed.

Done at the city of Washington this eleventh day of June in the year of our Lord one thousand nine hundred seventy-one and of the Independence of the United States of America the one hundred ninety-fifth.

Treaty for the Prohibition of Nuclear Weapons in Latin America

Preamble

In the name of their peoples and faithfully interpreting their desires and expirations, the Governments of the States which sign the Treaty for the Prohibition of Nuclear Weapons in Latin America,

Desiring to contribute, so far as lies in their power, towards ending the armaments race, especially in the field of nuclear weapons, and towards strengthening world at peace, based on the sovereign equality of States, mutual respect and neighbourliness.

Recalling that the United Nations General Assembly, in its Resolution 808 (X), adopted unanimously as one of three points of a coordinated programme disarmament "the total Prohibition of the use and manufacture of nuclear weapons and weapons of mass destruction of every type",

Recalling that militarily denuclearized zones are not an end in themselves but rather a means for achieving general and complete disarmament at a later stage,

Recalling United Nations General Assembly

Resolution 1911 (XVIII), which established that the measures that should be agreed upon for the denuclearization of Latin America should be taken "in the light of the principles of the charter of the United Nations and of regional agreements",

Recalling United Nations General Assembly Resolution 2028 (XX) which established the principle of an acceptable balance of mutual responsibilities for the nuclear and non-nuclear powers, and

Recalling that the Charter of the Organization of American States proclaims that it is an essential purpose of the Organization to strengthen the peace and parity of the hemisphere,

convinced:

That the incalculable destructive power of nuclear weapons has made it imperative that the legal prohibition of war should be strictly observed in practice if the survival of civilization and of mankind itself is to be assured.

That nuclear weapons, whose terrible effects are suffered, indiscriminately and exorably, by military force and civilian population alike, constitute, through persistence of the radioactivity they release, an attack on the integrity of human species and ultimately may even render the whole earth uninhabitable,

That general and complete disarmament under effective international control vital matter which all the peoples of the world equally demand,

That the proliferation of nuclear weapons, which seems inevitable unless States, the exercise of their sovereign rights, impose restrictions on themselves in or to prevent it, would make any agreement on disarmament enormously occult and would increase the danger of the outbreak of a nuclear conflagration,

That the establishment of militarily denuclearized zones is closely lined with maintenance of peace and security in the respective regions,

That the military denuclearization of vast geographical zones, adopted by the foreign decision of the States comprised therein, will exercise a beneficial reference on other regions where similar conditions exist,

That the privileged situation of the signatory States, whose territories are wholly free from nuclear weapons, imposes upon them the inescapable duty of preserving that situation both in their own interests and for the good of mankind,

That the existence of nuclear weapons in any country of Latin America would make it a target for possible nuclear attacks and would inevitably set off, throughout the region, a ruinous race in nuclear weapons which would involve the unjustifiable diversion, for warlike purposes, of the limited resources required for economic and social development.

That the foregoing reasons, together with the traditional peace loving outlook of Latin America, give rise to an inescapable necessity that nuclear energy should be used in that region exclusively for peaceful purposes, and that the Latin American countries should use their right to the greatest and most equitable possible access to this new source of energy in order to expedite the economic and social development of their peoples.

Convinced finally:

That the military denuclearization of Latin America being understood to mean the undertaking entered into internationally in this Treaty to keep their territories forever free from nuclear weapons—will constitute a measure which will spare their peoples from the squandering of their limited resources on nuclear armaments and will protect them against possible nuclear attacks on their territories, and will also constitute a significant contribution towards preventing the proliferation of nuclear weapons and a powerful factor for general and complete disarmament, and

That Latin America, faithful to its tradition of universality, must not only endeavour to banish from its homelands the scourge of a nuclear war, but must also strive to promote the well-being and advancement of its peoples, at the same time co-operating in the fulfilment of the ideals of mankind, that is to say, in the consolidation of a permanent peace based on equal rights, economic fairness, and social justice for all, in accordance with the principles and purposes set forth in the Charter of the United Nations and in the Charter of the Organization of American States,

Have agreed as follows:

Obligations

Article 1

1. The Contracting Parties hereby undertake to use exclusively for peaceful purposes the nuclear material and facilities which are under their jurisdiction, and to prohibit and prevent in their respective territories;

   (a) The testing use, manufacture, production or acquisition by any means whatsoever of any nuclear weapons, by the Parties themselves, directly or indirectly, on behalf of anyone else or in any other way,' and

   (b) The receipt, storage, installation, deployment and any form of possession of any nuclear weapons, directly or indirectly, by the Parties themselves, by anyone on their behalf or in any other way.

2. The Contracting Parties also undertake to refrain from engaging in, encouraging or authorizing,

directly or indirectly, or in any way participating in the testing, use, manufacture, production, possession or control of any nuclear weapon.

Definition of the Contracting Parties

Article 2

For the purposes of this Treaty, the Contracting Parties are those for whom the Treaty is in force.

Definition of Territory

Article 3

For the purposes of this Treaty, the term "territory shall include the territorial sea, air space and any other space over which the State exercises sovereignty in accordance with its own legislation.

Zone of application

Article 4

1. The zone of application of this Treaty is the whole of the territories for which the Treaty is in force.
2. Upon fulfilment of the requirements of article 28, paragraph 1, the zone of application of this Treaty shall also be that which is situated in the western hemisphere within the following limits (except the continental part of the territory of the United States of America and its territorial waters): starting at a point located at 35° north latitude, 75° west longitude; from this point directly southward to a point at 30° north latitude, 75° west longitude; from there, directly westward to a point at 30° north latitude, 50° west longitude; from there, along loxodromic line to a point at 5° north latitude, 20° west longitude; from there, directly southward to a point at 60° south latitude, 20° west longitude; from there, directly westward to a point at 60° south latitude; 115° west longitude; from there, directly

northward to point at 0 latitude, 115° west longitude; from there, along a loxodromic line to a point at 35° north latitude, 150° west longitude; from there, directly eastward to a point at 35° north latitude, 75° west longitude.

Definition of nuclear weapons

Article 5

For the purposes of this Treaty, a nuclear weapon is any device which is capable of releasing nuclear energy in an uncontrolled manner and which has group of characteristics that are appropriate for use for warlike purposes. An instrument that may be used for the transport or propulsion of the devise is not included in this definition if it is separable from the device and not an divisible part thereof.

Meeting of signatories

Article 6

At the request of any of the signatory States or if the Agency established by Article 7 should so decide, a meeting of all signatories may be convoked to consider in common questions which may affect the very essence of this instrument, including possible amendments to it. In either case, the meeting will be invoked by the General Secretary.

Organization

Article 7

1. In order to ensure compliance with the obligations of this Treaty, the Contracting Parties hereby establish an international organization to be known as the "Agency for the Prohibition of Nuclear Weapons in Latin America", hereinafter referred to as "the Agency". Only the Contracting Parties shall be affected by its decisions.

2. The Agency shall be responsible for the holding of periodic or extraordinary consultations among Member States on matters relating to the purposes, measures and procedures set forth in this Treaty and to the supervision of compliance with the obligations arising therefrom.
3. The Contracting Parties agree to extend to the Agency full and prompt co-operation in accordance with the provisions of the Treaty, of any agreements they may conclude with the Agency and of any agreements the Agency may conclude with any other international organization or body.
4. The headquarters of the Agency shall be in Mexico city.

Organs

Article 8

1. There are hereby established as principal organs of the Agency a General Conference, a Council and a Secretariat.
2. Such subsidiary organs as are considered necessary by the General Conference may be established within the purview of this Treaty.

The General Conference

Article 9

1. The General Conference, the supreme organ of the Agency, shall be composed of all the Contracting Parties; it shall hold regular sessions every two years, and may also hold special sessions whenever this Treaty so provides or, in the opinion of the Council, the circumstances so require.
2. The General Conference:

   (a) May consider and decide on any matters or questions covered by this Treaty, within the limits thereof, including those referring to powers and functions of any organ provided for in this Treaty.

(b) Shall establish procedures for the control system to ensure observance of this Treaty in accordance with its provisions.

(c) Shall elect the Members of the Council and the General Secretary.

(d) May remove the General Secretary from office if the proper functioning of the Agency so requires.

(e) Shall receive and consider the biennial and special reports submitted by the Council and the General Secretary.

(f) Shall initiate and consider studies designed to facilitate the optimum fulfilment of the aims of this Treaty, without prejudice to the power of the General Secretary independently to carry out similar studies for submission to and consideration by the Conference.

(g) Shall be the organ competent to authorize the conclusion of agreements with Governments and other international organizations and bodies.

3. The General Conference shall adopt the Agency's budget and fix the scale of financial, contribution to be paid by Member States, taking into account the systems and criteria used for the same purpose by the United Nations.

4. The General Conference shall elect its officers for each session and may establish such subsidiary organs as it deems necessary for the performance of its functions.

5. Each member of the Agency shall have one vote. The decisions of the General Conference shall be

taken by a two-thirds majority of the Members present and voting in the case of matters relating to the control system and measures referred to in article 20, the admission of new Members, the election or, removal of the General Secretary, adoption of the budget and matters related thereto. Decisions on other matters, as well as procedural questions and also determination of which question must be decided by a two-thirds majority, shall be taken by a simple majority of the Members present and voting.

6. The General Conference shall adopt its own rules of procedure.

The Council

Article 10

1. The Council shall be composed of five Members of the Agency elected by the General Conference from among the Contracting Parties, due account being taken of equitable geographic distribution.
2. The Members of the Council shall be elected for a term of four years. However, in the first election three will be elected for two years. Outgoing Members may not be re-elected for the following period unless the limited number of associates for which the Treaty is in force so requires.
3. Each member of the Council shall have one representative.
4. The Council shall be so organized as to be able to function continuously.
5. In addition to the functions conferred upon it by this Treaty and to those which may be assigned to it by the General Conference, the Council shall, through a General Secretary, ensure the proper operation of the control system in accordance with the provisions of

this Treaty and with the decisions adopted by the General Conference.

6. The Council shall submit an annual report on its work to the General Conference as well as such special reports as it deems necessary or which the general Conference requests of it.
7. The Council shall elect its officers for each session.
8. The decisions of the Council shall be taken by a simple majority of its members present and voting.
9. The Council shall adopt its own rules of procedure.

Secretariat

Article 11

1. The Secretariat shall consist of a General Secretary, who shall be the chief administrative officer of the Agency, and of such staff as the Agency may require. The term of office of the General Secretary shall be four years and may be re-elected for a single additional term. The General Secretary may be a national of the country in which the Agency has its headquarters. In the office of General Secretary becomes vacant, a new election shall be to fill the office for the remainder of the term.
2. The staff of the Secretarial shall be appointed by the General Secretary in accordance with rules laid down by the General Conference.
3. In addition to the functions conferred upon him by this Treaty and to those which may be assigned to him by the General Conference—the General Secretary shall ensure, as provided by article 10, paragraph 5, the proper operation of the control system established by this Treaty, in accordance with the provisions of the Treaty and the decisions taken by the General Conference.

4. The General Secretary shall act in that capacity in all meetings of the General Conference and of the Council and shall make an annual report to both bodies on the work of the Agency and any special reports requested by the General Conference or the Council or which the General Secretary may deem desirable.
5. The General Secretary shall establish the procedures for distributing to all Contracting Parties information received by the Agency from governmental sources and such information from non-governmental sources as may be of interest to the Agency.
6. In the performance of their duties the General Secretary and the staff shall not seek or receive instructions from any Government or from any other authority external to the Agency and shall refrain from any action which might reflect on their position as international officials responsible only to the Agency; subject to their responsibility to the Agency, they shall not disclose any industrial secrets or other confidential information coming to their knowledge by reason of their official duties in the Agency.
7. Each of the Contracting Parties undertakes to respect the exclusively international character of the responsibilities of the General Secretary and the staff and not to seek to influence them in the discharge of their responsibilities.

Control system

Article 12

1. For the purpose of verifying compliance with the obligations entered into by the Contracting Parties in accordance with article 1, a control system shall be established which shall be put into effect in accordance with the provisions of articles 13-18 of this Treaty.

2. The control system shall be used in particular for the purpose of verifying:
   (a) That devices, services and facilities intended for peaceful uses of nuclear energy are not used in the testing or manufacture of nuclear weapons,
   (b) That none of the activities prohibited in article 1 of this Treaty are carried out in the territory of the Contracting Parties with nuclear materials or weapons introduced from abroad, and
   (c) That explosions for peaceful purposes are compatible with article 18 of this Treaty.

IAEA safeguards

Article 13

Each Contracting Party shall negotiate multilateral or bilateral agreements with the International Atomic Energy Agency for the application of its safe—guards to its nuclear activities. Each Contracting Party shall initiate negotiations within a period of 180 days after the date of the deposit of its instrument of ratification of this Treaty. These agreements shall enter into force, for each Party, not later than eighteen months after the date of the initiation of such negotiations except in case of unforeseen circumstances or *force majeure.*

Reports of the Parties

Article 14

1. The Contracting Parties shall submit to the Agency and to the International Atomic Energy Agency, for their information, semi-annual reports stating that no activity prohibited under this Treaty has occurred in their respective territories.
2. The Contracting Parties shall simultaneously transmit to the Agency a copy of any report they may submit to the International Atomic Energy

Agency which relates to matters that are the subject of this Treaty and to the application of safeguards.

3. The Contracting Parties shall also transmit to the Organization of American States, for its information, any reports that may be of interest to it, in accordance with the obligations established by the Inter-American System.

Special reports requested by the General Secretary

Article 15

1. With the authorization of the Council, the General Secretary may request any of the Contracting Parties to provide the Agency with complementary or supplementary information regarding any event or circumstances connected with compliance with this Treaty, explaining his reasons. The Contracting Parties undertake to cooperate promptly and fully with the General Secretary.
2. The General Secretary shall inform the Council and the Contracting Parties forthwith of such requests and of the respective replies.

Special inspections

Article 16

1. The International Atomic Energy Agency and the Council established by its Treaty have the power of carrying out special inspections in the following cases:
   - (a) In the ease of International Atomic Energy Agency, in accordance with the agreements referred to in article 13 of this Treaty:
   - (b) In the case of the Council:
   - (i) When so requested, the reasons for the request being stated, by any Party which suspects that some activity prohibited by this Treaty has been

carried out or is about to be carried out, either in the territory of any other Party or in any other place on such latter Party's behalf, the Council shall immediately arrange for such an inspection in accordance with article 10, paragraph 5.

(ii) When requested by any Party which has been suspected of or charged with having violated this Treaty, the Council shall immediately arrange for the special inspection requested in accordance with article 10, paragraph 5.

The above requests will be made to the Council through the General Secretary.

The costs and expenses of any special inspection carried out under paragraph 1, sub-paragraph (b), sections (i) and (ii) of this article shall be borne the requesting Party or Parties, except where the Council concludes on the basis of the report on the special inspection that, in view of the circumstances existing in the case, such costs and expenses should be borne by the Agency.

3. The General Conference shall formulate the, procedures for the organization and execution of the special inspection carried out in accordance with paragraph 1, sub-paragraph (b), sections (i) and (ii) of this article.

4. The Contracting Parties undertake to grant the inspectors carrying out such special inspections full and free access to all places and all information which may be necessary for the performance of their duties and which are directly and intimately connected with the suspicion of violation of this Treaty. If so requested by the authorities of the Contracting Party in whose territory the inspection is carried out, the inspectors designated by the

General Conference shall be accompanied by representatives of said authorities, provided that this does not in any way delay or hinder the work of the inspectors.

5. The Council shall immediately transmit to all the Parties, through the General Secretary, a copy of any report resulting from special inspections.
6. Similarly, the Council shall sent through the General Secretary to the Secretary-General of the United Nations, for transmission to the United Nations Security Council and General Assembly, and to the Council of the Organization of American States, for its information, a copy of any report resulting from any special inspection carried out in accordance with paragraph 1, sub-paragraph (b), sections (i) and (ii) of this article.
7. The Council may decide, or any Contracting Party may request, the convening of a special session of the General Conference for the purpose of considering the reports resulting from any special inspection. In such a case, the General Secretary shall take immediate steps to convene the special session requested.
8. The General Conference, convened in special session under this article, may make recommendations to the Contracting Parties and submit reports to the Secretary-General of the United Nations to be transmitted to the United Nations Security Council and the General Assembly.

Use of nuclear energy for peaceful purposes

Article 17

Nothing in the provisions of this Treaty shall prejudice the rights of the Contracting Parties, in conformity with this Treaty, to use nuclear energy for peaceful purposes, in particular for their economic development and social progress.

Explosions for peaceful purposes

Article 18

1. The Contracting Parties may carry out explosions of nuclear devices for peaceful purposes-including explosions which involve devices similar to those used in nuclear weapons or collaborate with third parties for the same purpose, provided that they do so in accordance with the provisions of the article and the other articles of the Treaty, particularly articles 1 and 5.
2. Contracting Parties intending to carry out, or to cooperate in carrying out, such an explosion shall notify the Agency and the International Atomic Energy Agency as far in advance as the circumstances require, of the date of the explosion and shall at the same time provide the following information:
   (a) The nature of the nuclear device and the source from which it was obtained,
   (b) The place and purpose of the planned explosion,
   (c) The procedures which will be followed in order to comply with paragraph 8 of this article,
   (d) The expected force of the device, and
   (e) The fullest possible information on any possible radioactive fall-out that may result from the explosion or explosions, and measures which will be taken to avoid danger to the population, flora, fauna and territories of any other Party or Parties.
   (f) The General Secretary and the technical personnel designated by the Council and the

International Atomic Energy Agency may observe all the preparations, including the explosion of the device, and shall have unrestricted access to any in the vicinity of the site of the explosion in order to ascertain whether device and the procedures follows during the explosion are in conformity with the information supplied under paragraph 2 of this article and the other provisions of this Treaty.

The Contracting Parties may accept the collaboration of third parties for purpose set forth in paragraph 1 of the present article, in accordance with paragraphs 2 and 3 thereof.

Nations with other international organizations

Article 19

The Agency may conclude such agreements with the International Atomic Energy Agency as are authorized by the General Conference and as it considers by to facilitate the efficient operation of the control system established by this city.

The Agency may also enter into relations with any international organization or body, especially any which may be established in the future to supervise armaments or measures for the control of armaments in any part of the.

The Contracting Parties may, if they see fit, request the advice of the American Nuclear Energy Commission on all technical matters connected with the application of this Treaty with which the Commission is competent under its Statute.

Measures in the event of violation of the Treaty

Article 20

The General Conference shall take note of all cases in which, in its opinion, Contracting Party is not complying fully with its obligations under this treaty and shall draw the matter to the attention of the Party concerned, making recommendations as it deems appropriate.

If, in its opinion, such non-compliance constitutes a violation of this treaty which might endanger peace and security, the General Conference shall thereon simultaneously to the United Nations Security Council and the general Assembly through the Secretary-General of the United Nations, and to the Council of the Organization of American States. The General Conference shall likewise report to the International Atomic Energy Agency for such purposes as are relevant in accordance with its Statute.

United Nations and Organization of American States

Article 21

None of the provisions of this Treaty shall be construed as impairing the rights and obligations of the Parties under the Charter of the United Nations or, in the case of States Members of the Organization of American State, under existing regional treaties.

Privileges and immunities

Article 22

1. The Agency shall enjoy in the territory of each of the Contracting Parties such legal capacity and such privileges and immunities as may be necessary for the exercise of its functions and the fulfilment of its purposes.
2. Representatives of the Contracting Parties accredited to the Agency and officials of the Agency shall similarly enjoy such privileges and immunities as are necessary for the performance of their functions.
3. The Agency may conclude agreements with the contracting parties with a view to determining the

details of the application of paragraphs 1 and 2 of this article.

Notification of other agreements

Article 23

Once this Treaty has entered into force, the Secretarial shall be notified immediately of any international agreement concluded by any of the Contracting Parties on matters with which this Treaty is concerned; the Secretariat shall register, it and notify the other Contracting Parties.

Settlement of disputes

Article 24

Unless the Parties concerned agree on another mode of peaceful settlement, any question or dispute concerning the interpretation or application of this Treaty which is not settled shall be referred to the International Court of Justice with the prior consent of the Parties to the controversy.

Signature

Article 25

1. This Treaty shall be open indefinitely for signature by:
   (a) All the Latin American republics, and
   (b) All other sovereign States situated in their entirety south of latitude 35° north in the western hemisphere; and, except as provided in paragraph 2 of this article, all such States which become sovereign, when they have been admitted by the General Conference.
2. The General Conference shall not take any decision regarding the admission of a political entity part or all of whose territory is the subject, prior to the date

when this Treaty is opened for signature, or a dispute or claim between an extra-continental country and one or more Latin American States, so long as the dispute has not been settled by peaceful means.

Ratification and deposit

Article 26

1. This Treaty shall be subject to ratification by signatory States in accordance with their respective constitutional procedures.
2. This Treaty and the instruments or ratification shall be deposited with the Government of the Mexican United States, which is hereby designated the Depositary Government
3. The Depositary Government shall send certified copies of this Treaty to Governments of signatory States and shall notify them of the deposit of the instrument of ratification.

Reservations

Article 27

This Treaty shall not be subject to reservations.

Treaty into force

Article 28

Subject to the provisions of paragraph 2 of this article, this Treaty shall or into force among the States that have ratified it as soon as the following requirements have been met:

(a) Deposit of the instruments of ratification of this Treaty with the Depositary Government by the Governments of the States mentioned in article which are in existence on the date when this Treaty is opened for signature and which are not affected by the provisions of article 25, paragraph 2:

(b) Signature and ratification of Additional Protocol I annexed to this treaty by all extra-continental or continental States having *de jure* or *de facto* international responsibility for territories situated in the zone of application of the Treaty;

(c) Signature and ratification of the Additional Protocol II annexed to its Treaty by all powers possessing nuclear weapons;

(d) Conclusion of bilateral or multilateral agreements on the application the Safeguards System of the International Atomic Energy Agency in accordance with article 13 of this Treaty.

All signatory States shall have the imprescriptible right to waive, wholly part, the requirements laid down in the preceding paragraph. They may by means of a declaration which shall be annexed to their respective department of ratification and which may be formulated at the time of deposit instrument or subsequently. For those States which exercise this right, Treaty shall enter into force upon deposit of the declaration, or as soon whose requirements have been met which have not been expressly waived.

As soon as this Treaty has entered into force in accordance with the provision of paragraph 2 for eleven States, the Depositary Government shall convene preliminary meeting of those States in order that the Agency may be set up to commence its work.

After the entry into force of this Treaty for all the countries of the zone, of a new power possessing nuclear weapons shall have the effect of doing the execution of this Treaty for those countries which have ratified but waiving requirements of paragraph 1, sub-paragraph (c) of this article, which request such suspension: the Treaty shall remain suspended until the new power, on its own initiative or upon request by the General Conference, ratifies the annexed Additional Protocol II.

Amendments

Article 29

1. Any Contracting Party may propose amendments to this Treaty and shall submit its proposals to the Council through the General Secretary, who shall transmit them to all the other Contracting Parties and, in addition, to all other signatories in accordance with article 6. The Council, through the General Secretary, shall immediately following the meeting of signatories convene a special session of the General Conference to examine the proposals made, for the adoption of which a two-thirds majority of the Contracting Parties present and voting shall be required.
2. Amendments adopted shall enter into force as soon as the requirements set forth in article 28 of this Treaty have been complied with.

Duration and denunciation

Article 30

1. This Treaty shall be of a permanent nature and shall remain in force indefinitely, but any Party may denounce it by notifying the General Secretary of the Agency if, in the opinion of the denouncing State, there have arisen or may arise circumstances connected with the content of this Treaty or of the annexed Additional Protocols I and II which affect its supreme interests or the peace and security of one or more Contracting Parties.
2. The denunciation shall take effect three months after the delivery to the General Secretary of any Agency of the notification by the Government of the signatory State concerned. The General Secretary shall immediately communicate such notification to the other Contracting Parties and to the

Secretary-General of the United Nations for the information of the United Nations Security Council and the General Assembly. He shall also communicate it to the Secretary-General of the Organisation of American States.

Authentic Texts and Registration

Article 31

This Treaty, of which the Spanish, Chinese, English, French, Portuguese and Russian texts are equally authentic, shall be registered by the Depositary Government in accordance with article 102 of the United Nations Charter. The Depositary Government shall notify the Secretary-General of the Organization of American States for its information.

Transitional Article

Denunciation of the declaration referred to in article 28, paragraph 2, shall be subject to the same procedures as the denunciation of this Treaty, except that it will take effect on the date of delivery of the respective notification.

In witness whereof the undersigned Plenipotentiaries, having deposited their full powers, found in good and due form, sign this Treaty on behalf of their respective Governments.

Done at Mexico, Distrito Federal, on the Fourteenth day of February, one thousand nine hundred and sixty-seven.

Additional Protocol I

The undersigned Plenipotentiaries, furnished with full powers by their respective Governments,

Convinced that the Treaty for the Prohibition of Nuclear Weapons in Latin America, negotiated and signed in accordance with the recommendations of

General Assembly of the United Nations in Resolution 1911 (XVIII) of 27 November 1963, represents an important step towards ensuring the non-proliferation of nuclear weapons,

Aware that the non-proliferation of nuclear weapons is not an end in itself, rather, a means of achieving general and complete disarmament at a later stage, and

Desiring to contribute, so far as lies in their power, towards ending the armaments race, especially in the field of nuclear weapons, and towards strengthening world at peace, based on mutual respect and sovereign equality of States,

We agreed as follows:

Article 1.

To undertake to apply the statute of denuclearization in respect of warlike purposes as defined in articles 1, 3, 5 and 13 of the Treaty for the Prohibition of Nuclear Weapons in Latin America in territories for which, de jure de facto, they are internationally responsible and which lie within the limits the geographical zone established in the Treaty.

Article 2.

The duration of the Protocol shall be the same as that of the Treaty the Prohibition of Nuclear Weapons in Latin America of which this Protocol an annex, and the provisions regarding ratification and denunciation contained the Treaty shall be applicable to it.

Article 3.

This Protocol shall enter into force, for the States which have ratified it, on the date of the deposit of their respective instruments of ratification.

In witness whereof the undersigned Plenipotentiaries, having deposited their all powers, found in good and due form sign this Protocol on behalf of their respective Governments.

The United Kingdom and the Netherlands are parties to this Protocol.

Treaty for the Prohibition of Nuclear Weapons in Latin America. Done at Mexico February 14, 1967

Country	Date of Signature	Date of Deposit of Ratification
Argentina	9/27/67	
Barbados	10/18/68	4/25/69
Bolivia	2/14/67	2/18/69
Brazil	5/9/67	1/29/68
Chile	2/14/67	10/9/74
Colombia	2/14/67	8/4/72
Costa rica	2/14/67	8/25/69
Dominican Republic	7/29/67	6/14/68
Ecuador	2/14/67	2/11/69
El Salvador	2/14/67	4/22/68
Guatemala	2/14/67	2/6/70
Haiti	2/14/67	5/23/69
Honduras	2/14/67	9/23/68
Jamaica	10/26/67	6/26/69
Mexico	2/14/67	9/20/67
Nicaragua	2/15/67	10/24/68
Panama	2/14/67	6/11/71
Paraguay	4/26/67	3/19/69
Peru	2/14/67	3/4/69
Trinidad & Tobago	6/27/67	12/3/70

Uruguay	2/14/67	8/20/68
Venezuela	2/14/67	3/23/70
Total	22	19

National Protocol I to the Treaty of February 14, 1967 the Prohibition of Nuclear Weapons in Latin America. Done at Mexico February 14, 1967.

Country	Date of Signature	Date of Deposit of Ratification
Netherlands	4/1/68	7/26/71
United Kingdom	12/20/67	12/11/69

National Protocol II to the Treaty of February 14, for the Prohibition of Nuclear Weapons in Latin America. Done at Mexico February 14, 1967.

Country	Date of Date of Signature	Deposit of Ratification
Kingdom	12/20/67	12/11/69
States	4/1/68	5/12/71
People's Republic of	8/21/73	6/12/74
	7/18/73	3/22/74

10 Convention on Third Party Liability in the Field of Nuclear Energy

Paris, 29 July 1960

The Government of the Federal Republic of Germany, the Republic of Austria, the Kingdom of Belgium, the Kingdom of Denmark, Spain, the French Republic, the Kingdom of Greece, the Italian Republic, the Grand Duchy of Luxembourg, the Kingdom of Norway, the Kingdom of the Netherlands, the Portuguese Republic, the United Kingdom of Great Britain and Northern Ireland, the Kingdom of Sweden, the Swiss Confederation and the Turkish Republic;

Considering that the European Nuclear Energy Agency, established within the framework of the Organization for European Economic co-operation (hereinafter referred to as the "Organization"), is charged with encouraging the elaboration and humanization of legislation relating to nuclear energy in participating countries, in particular with regard to third party liability and insurance against atomic risks;

Desirous of ensuring adequate and equitable compensation for persons who suffer damage caused by nuclear incidents whilst taking the necessary steps to ensure that the development of the production and uses of nuclear energy for peaceful purposes is not, thereby hindered;

Convinced of the need for unifying the basic rules applying in the various countries to the liability incurred for such damage, whilst leaving these countries free to take, on a national basis, any additional measures which they deem appropriate, including the application of the provisions of this Convention to damage caused by incidents due to ionizing radiations not covered therein;

Have agreed as follows:

Article 1

a) For the purposes of this Convention:

(i) "A nuclear incident" means any occurrence or succession of occurrences having the same origin which cause damage, provided that such occurrence or succession of occurrences, or any of the damage caused, arises out of or results from the radioactive properties, or a combination of radioactive properties with toxic, explosive, or other hazardous properties of nuclear fuel or radioactive products or waste or with any to them.

(ii) "Nuclear installation" means reactors other than those comprised in any means or transport; factories for the manufacture or processing of nuclear substances; factories for the separation of isotopes of nuclear fuel; factories for the reprocessing of irradiated nuclear fuel; facilities for the storage of nuclear substances other than storage incidental to the carriage of such substances; and such other installations in which there are nuclear fuel or radioactive products or waste as the Steering Committee of the European Nuclear Agency (hereinafter referred to as the "Steering Committee") shall from time to time determine.

(iii) "Nuclear fuel" means fissionable material in the form of uranium metal, alloy, or chemical compound (including natural uranium), plutonium metal, alloy, or compound, and such other fissionable material as the Steering Committee shall from time to time determine.

(iv) "Radioactive products or waste" means any radioactive material produced in or made radioactive by exposure to the radiation incidental to the process of producing or utilizing nuclear fuel, but does not include (1) nuclear fuel, or (2) radioisotopes outside a nuclear installation which are used or intended to be used for any industrial, commercial, agricultural, medical or scientific purpose.

(v) "Nuclear substances" means nuclear fuel (other than natural uranium and other than depleted uranium) and radioactive products or waste.

(vi) "Operator" in relation to a nuclear installation means the person designated or recognized by the competent public authority as the operator of that installation.

b) The Steering Committee may, if in its view the small extent of the risks involved so warrants, exclude any nuclear installation, nuclear fuel or nuclear substances from the application of this Convention.

Article 2

This Convention does not apply to nuclear incidents occurring in the territory of non-Contracting States or to damage suffered in such territory, unless otherwise provided by the legislation of the Contracting Party in whose territory the nuclear installation of the operator liable is situated, and except in regard to rights referred to in Article 6(e).

Article 3

a) The operator of a nuclear installation shall be liable, in accordance with this Convention, for:

(i) damage to or loss of life of any person; and

(ii) damage to or loss of any property other

1. The nuclear installation itself and any property on the site of that installation which is used or to be used in connection with that installation:

2. In the cases within Article 4, the means of transport upon which the nuclear substances involved were at the time of the nuclear incident, upon proof that such damage or loss (hereinafter referred to as "damage") was caused by a nuclear incident involving either nuclear fuel or radioactive products or waste in, or nuclear substances coming from such installation, except as otherwise provided for in Article 4.

b) Where the damage or loss is caused jointly by a nuclear incident and by an incident other than a nuclear incident, that part of the damage or loss which is caused by such other incident shall, to the extent that it is not reasonably separable from the damage or loss caused by the nuclear incident, be considered to be damage caused by the nuclear incident. Where the damage or loss is caused jointly by a nuclear incident and by an emission of ionizing radiation not covered by this Convention, nothing in this Convention shall limit or otherwise affect the liability of any person in connection with that emission of ionizing radiation.

c) Any Contracting Party may by legislation provide that the liability of the operator of a nuclear installation situated in its territory shall include liability for damage which arises out of or results from ionizing radiations emitted by any source of radiation inside that installation, other than those referred to in paragraph (a) of this Article.

Article 4

In the case of carriage of nuclear substances, including storage incidental thereto, without prejudice to Article 2:

a) The operator of a nuclear installation shall be liable, in accordance with this Convention, for damage upon proof that it was caused by a nuclear incident outside that installation and involving nuclear substances in the course of carriage therefrom, only if the incident occurs:

   (i) before liability with regard to nuclear incidents involving the nuclear substances has been assumed, pursuant to the express terms of a contract in writing, by the operator of another nuclear installation;

   (ii) in the absence of such express terms, before the operator of another nuclear installation has taken charge of the nuclear substances; or

   (iii) where the nuclear substances are intended to be used in a reactor comprised in a means of transport, before the person duly authorized to operate that reactor has taken charge of the nuclear substances; but

   (iv) where the nuclear substances have been sent to a person within the territory of a non-Contracting State, before they have been unloaded from the means of transport by which they have arrived in the territory of that non-Contracting State.

b) The operator of a nuclear installation shall be liable, in accordance with this Convention, for damage upon proof that it was caused by a nuclear incident outside that installation and involving nuclear substances in the course of carriage thereto, only if the incident occurs:

(i) after liability with regard to nuclear incidents involving the nuclear substances has been assumed by him, pursuant to the express terms of a contract in writing, from the operator of another nuclear installation;

(ii) in the absence of such express terms, after he has taken charge of the nuclear substances; or

(iii) after he has taken charge of the nuclear substances from a person operating a reactor comprised in a means of transport; but

(iv) where the nuclear substances have, with the written consent of the operator, been sent from a person within the territory of a non-Contracting State, after they have been loaded on the means of transport by which they are to be carried from the territory of that State.

c) The operator liable in accordance with this Convention shall provide the carrier with a certificate issued by or on behalf of the insurer or other financial guarantor furnishing the security required pursuant to Article 10. The certificate shall state the name and address of that operator and the amount, type and duration of the security, and these statements may not be disputed by the person by whom or on whose behalf the certificate was issued. The certificate shall also indicate the nuclear substances and the carriage in respect of which the security applies and shall include a statement by the competent public authority that the person named is an operator within the meaning of this Convention.

d) A Contracting Party may provide by legislation that, under such terms as maybe contained therein and upon fulfilment of the requirements of Article 10(a), a carrier may, at his request and with the consent of an operator of a nuclear installation situated in its territory, by decision of the competent public authority, be liable in accordance with this Convention in place of that operator. In such a case for all the purposes of this Convention the carrier shall be considered, in respect of nuclear installation on the territory of the Contracting Party whose legislation so provides.

Article 5

a) If the nuclear fuel or radioactive products or waste involved in a nuclear incident have been in more than one nuclear installation and are in a nuclear installation at the time damage is caused, no operator of any nuclear installation in which they have previously been shall be liable for the damage.

b) Where, however, damage is caused by a nuclear incident occurring in a nuclear installation and involving only nuclear substances stored therein incidentally to their carriage, the operator of the nuclear installation shall not be liable where another operator or person is liable pursuant to Article 4.

c) If the nuclear fuel or radioactive products or waste involved in a nuclear incident have been in more than one nuclear installation at the time damage is caused, no operator other than the operator of the last nuclear installation in which they were before the damage was caused or and operator who has subsequently taken them in•charge shall be liable for the damage.

d) If damage gives rise to liability of more than one operator in accordance with this Convention, the liability of these operators shall be joint and several: provided that where such liability arises as a result of damage caused by a nuclear incident involving nuclear substances in the course of carriage in one and the same means of transport, or, in the case of storage incidental to the carriage, in one and the same nuclear installation, the maximum total amount for which such operators shall be liable shall be the highest amount established with respect to any of them pursuant to Article 7 and provided that in no case shall any one operator be required, in respect, of a nuclear incident, to pay more than the amount established with respect to him pursuant to Article 7.

Article 6

(a) The right to compensation for damage caused by a nuclear incident may be exercised only against an operator liable for the damage in accordance with this Convention, or, if a direct right of action against the insurer or other financial guarantor furnishing the security required pursuant to Article 10 is given by national law, against the insurer or other financial guarantor.

(b) Except as otherwise provided in this Article, no other person shall be liable for damage caused by nuclear incident, by this provision shall not affect the application of any international agreement in the field of transport in force or open for signature, ratification or accession at the date of this Convention.

(c) (i) Nothing in this Convention shall affect the liability:

1. Of any individual for damage caused by a nuclear incident for which the operator, by virtue of Articles 3 (a)(ii)(1) and (2) or article 9, is not liable under this Convention and which results from an act or omission of that individual done with intent to cause damage;

2. Of a person duly authorised to Operate a reactor comprised in a means of transport for damage caused by a nuclear incident when an operator is not liable for such damage pursuant to Article 4(a)(iii) or (b)(iii).

   (ii) The operator shall incur no liability outside this Convention for damage caused by a nuclear incident except where use has not been made of the right provided for in Article 7(c), and then only to the extent that national legislation or the legislation of the Contracting Party in whose territory in nuclear installation of the operator liable is situated has made specific provisions concerning damage to the means of transport.

(d) Any person who has paid compensation in respect of damage caused by a nuclear incident under any international agreement referred to in Paragraph (b) of this Article or under any legislation of a non-Contracting State shall, up to the amount which he has paid, acquire by subrogation the rights under this Convention of the person suffering damage whom he has so compensated.

(e) Any person who has his principal place of business in the territory of a Contracting Party or who is the servant of such a person and who has paid compensation in respect of damage caused by a nuclear incident occurring in the territory of a non-Contracting State or in respect of damage suffered in such territory shall, up to the amount which he has paid, acquire the right which the person so compensated would have had against the operator but for the provisions of article 2,

(f) The operator shall have a right of recourse only:

   (i) if the damage caused by a nuclear incident

results from an act or omission done with intent to cause damage, against the individual acting or omitting to act with such intent;

(ii) if and to the extent that it is so provided expressly by the contract.

(g) If the operator has a right of recourse to any extent pursuant to paragraph (f) of this Article against any person, that person shall not, to that extent, have a right against the operator under paragraphs (d) or (e) of this Article.

(h) Where provisions of national or public health insurance, social security, workmen's compensation or occupational disease compensation systems include compensation for damage caused by a nuclear incident, rights of beneficiaries of such systems and rights of recourse by virtue of such systems shall be determined by the law of the Contracting Party or by the regulations of the inter-Governmental organization which has established such systems.

Article 7

a) The aggregate of compensation required to be paid in respect of damage caused by a nuclear incident shall not exceed the maximum liability established in accordance with this Article.

b) The maximum liability of the operator in respect of damage caused by a nuclear incident shall be 15,000,000 European Monetary Agreement units of account as defined at the date of this Convention (hereinafter referred to as "units of account") provided that any Contracting Party, taking into account the possibilities for the operator of obtaining the insurance or other financial security required pursuant to Article 10, may establish by legislation a greater or less amount but in no event less that

5,000,000 units of account. The sums mentioned above may be converted into national currency in round figures.

c) Any Contracting Party may by legislation provide that the exception in Article 3(a)(ii)(2) shall not apply: provided that in no case shall the inclusion of damage to the means of transport result in reducing the liability of the operator in respect of other damage to, an amount less than 5,000,000 units of account.

d) The amount of liability of operators of nuclear installations in the territory of a Contracting Party established in accordance with paragraph (b) of this Article as well as the provisions of any legislation of a Contracting Party pursuant to paragraph (c) of this article shall apply to the liability of such operators wherever the nuclear incident occurs.

e) A Contracting Party may subject the transit of nuclear substances through its territory to the condition that the maximum amount of liability of the foreign operator concerned be increased, if it considers that such amount does not adequately cover the risks of a nuclear incident in the course of the transit: provided that the maximum amount thus increased shall not exceed the maximum amount of liability of operators of nuclear installations situated in its territory.

f) The provisions of paragraph (e) of this Article shall not apply:

(i) to carriage by sea where, under international law, there is a right of entry in cases of urgent distress into the ports of such Contracting Party or a right of innocent passage through its territory.

(ii) to carriage by air where, by agreement or under

international law there is a right to fly over or land on the territory of such Contracting Party.

g) Any interest and costs awarded by a court in actions for compensation under this Convention shall not be compensation for the purposes of this Convention and shall be payable by the operator in addition to any sum for which he is liable in accordance with this Article.

Article 8

a) The right of compensation under this Convention shall be extinguished if an action is not brought within ten years from the date of the nuclear incident. National legislation may, however, establish a period longer than ten years if measures have been taken by the Contracting Party in whose territory the nuclear installation of the operator liable is situated to cover the liability for that operator in respect of any actions for compensation begun after the expiry of the period of ten years and during such longer period: provided that such extension of the extinction period shall in no case affect the right of compensation under this Convention of any person who has brought an action in respect of loss of life or personal injury against the operator before the expiry of the period of ten years.

b) In the case of damage caused by a nuclear incident involving nuclear fuel or radioactive products or waste which, at the time of the incident have been stolen, lost jettisoned or abandoned and have not yet been recovered, the period established pursuant to paragraph (a) of this Article shall be computed from the date of that nuclear incident, but the period shall in no case exceed twenty years from the date of the theft, loss, jettison or abandonment.

c) National legislation may establish a period of not less than less than tow years from the extinction of the right or as a period of limitation either from the date at which the person suffering damage has knowledge or from the date at which he ought reasonably to have known of both the damage and the operator liable; provided that the period established pursuant to paragraph (a) and (b) of this Article shall not be exceeded.

d) Where the provisions of Article 13(c)(ii) are applicable, the right of compensation shall not, however, be extinguished if, within the time provided for in paragraph (a) of this Article,

(i) prior to the determination by the Tribunal referred to in Article 17, an action has been brought before any of the courts from which the Tribunal can choose; if the Tribunal determines that the competent court is a court other than that before which such action has already been brought, it may fix a date by which such action has to be brought before the competent court so determined; or

(ii) a request has been made to a Contracting Party concerned to initiate a determination by the Tribunal of the competent court pursuant to Article 13 (c) (ii) and an action is brought subsequent to such determination within such time as may be fixed by the Tribunal.

e) Unless national law provides to the contrary, any person suffering damage caused by a nuclear incident who has brought an action for compensation within the period provided for in this Article may amend his Claim in respect of any aggravation of the damage after the expiry of such period provided that final judgement has not been entered by the competent court.

Article 9

The operator shall not be liable for damage caused by a nuclear incident directly due to an act of armed conflict, hostilities, civil war, insurrection or except in so far as the legislation of the Contracting Party in whose territory his nuclear installation is situated may provide to the contrary, a grave natural disaster of an exceptional character.

Article 10

a) To cover the liability under this Convention, the operator shall be required to have and maintain insurance or other financial security of the amount established pursuant to Article 7 and of such type and terms as the competent public authority shall specify.

b) No insurer or other financial guarantor shall suspend or cancel the insurance or other financial security provided for in paragraph (a) of this Article without giving notice in writing of at least two months to the competent public authority or in so far as such insurance or other financial security relates to the carriage of nuclear substances, during the period of the carriage in question.

c) The sums provided as insurance, reinsurance or other financial security may be drawn upon only for compensation for damage caused by a nuclear incident.

Article 11

The nature from and extent of the compensation, with the limits of this Convention, as well as the equitable distribution thereof, shall be governed by national law.

Article 12

Compensation payable under this Convention, insurance and reinsurance premiums, sums provided as insurance, reinsurance, or other financial security required pursuant to Article 10, and interest and costs referred to in Article 7(g), shall be freely transferable between the monetary areas of the Contracting Parties.

Article 13

a) Except as otherwise provided in this Article, jurisdiction over actions under Articles 3, 4, 6(a) and 6(e) shall lie only with the courts of the Contracting Party in whose territory the nuclear incident occurred.

b) Where a nuclear incident occurs outside the territory of the Contracting Parties, or where the place of the nuclear incident cannot be determined with certainty, jurisdiction over such actions shall lie with the courts of the Contracting Parties in whose territory the nuclear installation of the operator liable is situated.

c) Where jurisdiction would lie with the courts of more than one Contracting Party by virtue of paragraphs (a) or (b) of this Article, jurisdiction shall lie,

   (i) if the nuclear incident occurred partly outside the territory of any Contracting Party and partly in the territory of single Contracting Party, with the courts of that Contracting Party; and

   (ii) in any other case, with the courts of the Contracting Party determined, at the request of a Contracting Party concerned, by the Tribunal referred to in Article 17 as being the most closely related to the case in question.

d) Judgements entered by the competent court under this Article after trial, or by default, shall, when they have become enforceable under the law applied by that court, become enforceable in the territory of any of the other Contracting Parties as soon as the formalities required by the Contracting Party concerned have been complied with. The merits of the case shall not be the subject of further proceedings. The foregoing provisions shall not apply to interim judgements.

e) If an action is brought against a Contracting Party under this Convention, such Contracting Party may not, except in respect of measures of execution, invoke any jurisdictional immunities before the court competent in accordance with this Article.

Article 14

a) This Convention shall be applied without any discrimination based upon nationality, domicile, or residence.

b) "National Law" and "National Legislation" mean the national law or the national legislation of the court having jurisdiction under this Convention over claims arising out of a nuclear incident, and that law or legislation shall apply to all matters both substantive and procedural not specifically governed by this Convention.

c) That law and legislation shall be applied without any discrimination based upon nationality, domicile, or residence.

Article 15

a) Any Contracting Party may take such measures as it deems necessary to provide for an increase in the amount of compensation specified in this Convention.

(b) In so far as compensation for damage involves public funds and is in excess of the 5,000,000 units of account referred to in Article 7, any such measures in whatever form may be applied under conditions which may derogate from the provisions of this Convention

Article 16

Decisions taken by the Steering Committee under Article 1 (a) (ii) (a) (iii) and 1 (b) shall be adopted by mutual agreement of the members representing the Contracting Parties.

Article 17

Any dispute arising between two or more Contracting Parties concerning the interpretation or application this Convention shall be examined by the Steering Committee and in the absence of friendly settlement shall, upon the request of Contracting Party concerned, be submitted to the Tribunal established by the Convention of 20 December 1957 on the Establishment of Security Control in the Field of Nuclear Energy.

Article 18

a) Reservations to one or more of the provisions of this Convention may be made at any time prior to ratification of or accession to this Convention or prior to the time of notification under Article 23 in respect of any territory or territories mentioned in the notification, and shall be admissible only if the terms of these reservations have been expressly accepted by Signatories.

b) Such acceptance shall not be required from a Signatory which has not itself ratified this Convention within a period of twelve months after the date of notification to it of such reservation by the Secretary-General of the Organization in accordance with Article 24.

c) Any reservation admitted in accordance with this Article may be withdrawn at any time by notification addressed to the Secretary-General of the Organization.

Article 19

a) This Convention shall be ratified. Instruments of ratification shall be deposited with the Secretary-General of the Organization.

b) This Convention shall come into force upon the deposit of instruments of ratification by not less than five of the Signatories. For each Signatory ratifying thereafter, this Convention shall come into force upon the deposit of its instrument of ratification.

Article 20

Amendments to this Convention shall be adopted by mutual agreement of all the Contracting Parties. They shall come into force when ratified or confirmed by two-thirds of the Contracting Parties. For each Contracting Party ratifying or confirming thereafter, they shall come into force at the date of such ratification or confirmation.

Article 21

a) The Government of any Member or Associate country of the Organization which is not a Signatory to this Convention may accede thereto by notification addressed to the Secretary-General of the Organization.

b) The Government of any other country which is not a Signatory to this Convention may accede thereto by notification addressed to the Secretary-General of the Organization and with the unanimous assent of the Contracting Parties. Such accession shall take effect from the date of such assent.

Article 22

a) This Convention shall remain in effect for a period of ten years as from the date of its coming into force. Any Contracting Party may, by giving twelve months' notice to the Secretary-General of the Organization, terminate the application of this Convention to itself at the end of the period of ten years.

b) This Convention shall, after the period of ten years, remain in force for a period of five years for such Contracting Parties as have not terminated its application in accordance with paragraph (a) of this Article, and thereafter for successive periods of five years for such Contracting Parties as have not terminated its application at the end of one of such periods of five years by giving twelve months' notice to that effect to the Secretary-General of the Organization.

c) A conference shall be convened by the Secretary-General of the Organization in order to consider revisions to this Convention after a period of five years as from the date of its coming into force or, at any other time, at the request of a Contracting Party, within six months from the date of such request.

Article 23

a) This Convention shall apply to the metropolitan territories of the Contracting Parties.

b) Any Signatory or Contracting Party may, at the time of signature or ratification of or accession to this Convention or at any later time, notify the Secretary-General of the Organization that this Convention shall apply to those of its territories, including the territories for whose international relations it is responsible, to which this Convention is not applicable in accordance with paragraph (a) of this Article and which are mentioned in the notification. Any such notification may in respect of any territory or territories mentioned therein be withdrawn by giving twelve months' notice to that effect to the Secretary-General of the Organization.

c) Any territories of Contracting Party, including the territories for whose international relations it is responsible, to which this Convention does not apply shall be regarded for the purposes of this Convention as being a territory of a non-Contracting State.

Article 24

The Secretary-General of the Organization shall give notice to all Signatories and acceding Governments of the receipt of any instrument of ratification, accession, withdrawal, notification under Article 23, and decisions of the Steering Committee under Article 1(a)(ii), 1(a)(iii) and 1(b). He shall also notify them of the date on which this Convention comes into force, the text of any amendment thereto and of the date on which such amendment comes into force, and any reservation made in accordance with Article 18.

In Witness Whereof, the undersigned Plenipotentiaries, duly empowered, have signed this Convention.

Done in Paris, this twenty-ninth day of July Nineteen Hundred and Sixty in the English, French, German, Spanish, Italian and Dutch languages in a single copy which shall remain deposited with the Secretary-General of the Organization for European Economic Co-operation by whom certified copies will be communicated to all Signatories.

Annex 1

The following reservations were accepted either at the time of signature of the Convention or at the time of signature of the Additional Protocol:

1. Article 6(a) and (c) (i)

Reservation by the Government of the Federal Republic of Germany, the Government of the Republic of Austria and the Government of the Kingdom of Greece.

Reservation of the right to provide, by national law, that persons other than the operator may continue to be liable for damage caused by nuclear incident on condition that these persons are fully covered in respect of their liability, including defence against unjustified actions, by insurance or other financial security obtained by the operator or out of State funds.

2. Article 6(b) and (d)

Reservation by the Government of the Republic of Austria, the Government of Kingdom of Greece, the Government of Kingdom of Norway and the Government of Kingdom of Sweden.

Reservation of the right to consider their national legislation which includes provisions equivalent to those included in the international agreements referred to in Article 6(b) as being international agreements within the meaning of Article 6(b) and (d).

3. Article 8 (a)

Reservation by the Government of the Federal Republic of Germany and the Government of the Republic of Austria.

Reservation of the right to establish, in respect of nuclear incidents occurring in the Federal Republic of Germany and in the Republic of Austria respectively, a period longer than ten years if measures have been taken to cover the liability of the operator in respect of any actions for compensation begun after the expiry of the period of ten years and during such longer.

4. Article 9

Reservation by the Government of Federal Republic of. Germany and the Government of the Republic of Austria.

Reservation of the right to provide, in respect of nuclear incidents occurring in the Federal Republic of Germany and the Republic of Austria respectively, that the operator shall be liable for damage caused by a nuclear incident directly due to an act of armed conflict, hostilities, civil war, insurrection or a grave natural disaster of an exceptional character.

5. Article 19

Reservation by the Government of the Federal Republic of Germany, the Government of the Republic of Austria, and the Government of the Kingdom of Greece.

Reservation of the right to consider ratification of this Convention as constituting an obligation under international law to enact national legislation on third party liability in the field of nuclear energy in accordance with the provisions of this Convention.

11 Convention Supplementary to the Paris Convention of 29 July 1960 on Third Party Liability in the Field of Nuclear Enrgy (1963)

Brussels, 31 January 1963

The Governments of the Federal Republic of Germany, the Republic of Austria, the Kingdom of Belgium, the Kingdom of Denmark, Spain, the French Republic, the Italian Republic, the Grand Duchy of Luxembourg, the Kingdom of Norway, the Kingdom of the Netherlands, the United Kingdom of Great Britain, and Northern Ireland, the Kingdom of Sweden and the Swiss Confederation,

Being Parties to the Convention of 29 July 1960 on Third Party Liability in the Field of Nuclear Energy, concluded within the framework of the Organization for European- Economic Co-operation and Development (hereinafter referred to as the "Paris Convention"),

Desirous of supplementing the, measures provided in that Convention with a view to increasing the amount of compensation for damage which might result from the use of nuclear energy for peaceful purposes,

Have agreed as follows:

Article 1

The system instituted by this Convention is supplementary to that of the Paris Convention, shall be subject to the provisions of the Paris Convention, and shall be applied in accordance with the following Articles.

Article 2

a) The system of this Convention shall apply to damage caused by nuclear incidents, other than those occurring entirely in the territory of a State which is not a Partly to this Convention:

(i) for which an operator of a nuclear installation, used for peaceful purposes, situated in the territory of a Contracting Party to this Convention, (hereinafter referred to as a "Contracting Party"), and which appears on the list established and kept up-to-date in accordance with the terms of Article 13, is liable under the Paris Convention;

and

(ii) suffered

1. In the territory of Contracting Party, or
2. On or over the high seas on board a ship or aircraft registered in the territory of a Contracting Party;
3. On or over the high seas by a national of a Contracting Party, provided that, in the case of damage to a ship or an aircraft, the ship or aircraft is registered in the territory of a Contracting Party, provided that the courts of a Contracting Party have jurisdiction pursuant to the Paris Convention.

b) Any Signatory or acceding Government may, at the time of signature of or accession to this Convention or on the deposit of its instrument of ratification, declare that, for the purpose of the application of paragraph (a)(ii) of this Article, individuals of certain categories thereof, considered under its law as having their habitual residence in its territory, are assimilated to its own nationals.

c) In this Article the expression "a national of a Contracting Party" shall include a Contracting Party or any of its constituent sub-divisions, or a partnership, or any public or private body whether corporate or not established in the territory of a Contracting Party.

Article 3

a) Under the conditions established by this Convention, the Contracting Parties undertake that compensation in respect of the damage referred to in Article 2 shall be provided up to the amount of 120 million units of account per incident.

b) Such compensation shall be provided:

   (i) up to an amount of at least 5 million units of account, out of funds provided by insurance or other financial security, such amount to be established by the legislation of the Contracting Party in whose territory the nuclear installation of the operator liable is situated;

   (ii) between this amount and 70 million units of accounts, out of public funds to be made available by the Contracting Party in whose territory the nuclear installation of the operator liable is situated:

   (iii) between 70 million and 120 million units of account, out of public funds to be made available by the Contracting Parties according to the formula for contributions specified in Article 12.

c) For this purpose, each Contracting Party shall either:

(i) establish the maximum liability of the operator, pursuant to Article 7 of the Paris Convention, at 120 million units of account, and provide that such liability shall be covered by all the funds referred to in paragraph (b) of this Article; or

(ii) establish the maximum liability of the operator at an amount at least equal to that established pursuant to paragraph (b)(i) of this Article and provide that, in excess of such amount and up to 120 million units of account, the public funds referred to in paragraph (b)(ii) and (iii) of this Article shall be made available by some means other than as cover for the liability of the operator, provided that the rules of substance and procedure laid down in this Convention are not thereby affected.

d) The obligation of the operator to pay compensation, interest or costs out of public funds made available pursuant to paragraph (b)(ii) and (iii) and (f) of this Article shall only be enforceable against the operator as and when such funds are in fact made available.

e) The Contracting Parties, in carrying out this Convention, undertake not to make use of the right provided for in Article 15(b) of the Paris Convention to apply special conditions.

(i) in respect of compensation for damage provided out of the funds referred to in paragraph (b)(i) of this Article;

(ii) other than those laid down in this Convention in respect of compensation for damage provided out of the public funds referred to in paragraph (b)(ii) and (iii) of this Article.

f) The interest and costs referred to in Article 7 (g) of the Paris Convention are payable in addition to the amounts referred to in paragraph (b) of this Article and shall be borne in so far as they are awarded in respect of compensation payable out of the funds referred to in:

(i) paragraph (b)(i) of this Article, by the operator liable;

(ii) paragraph (b)(ii) of this Article, by the Contracting Party in whose territory the nuclear installation of that operator is situated;

(iii) paragraph (b)(iii) of this Article by the Contracting Parties together.

g) For the purpose of this Convention, "unit of account" means the unit of account of the European Monetary Agreement as defined at the date of the Paris Convention.

Article 4

a) If a nuclear incident causes damage which gives rise to liability of more than one operator, the aggregate liability provided for in Article 5 (b) of the Paris Convention shall not, to the extent that public funds have to be made available pursuant to Article 3 (b)(ii) and (iii), exceed 120 million units of account.

b) The total amount of the public funds made available pursuant to Article 3(b)(ii) and (iii) shall not, in such event, exceed the difference between 120 million units of account and the sum of the amounts established with respect to such operators pursuant to Article 3(b)(i) or, in the case of an operator whose nuclear installation is situated in the territory of a State which is not a Party to this Convention, the amount established pursuant to Article 7 of the Paris Convention. If more than one Contracting, Party is required to make available public funds pursuant to Article 3(b)(ii), such funds shall be made available by them in proportion to the number of nuclear installations situated in their respective territories, which are involved in the nuclear incident and of which the operators are liable.

Article 5

a) Where the operator liable has a right of recourse pursuant to Article 6(f) of the Paris Convention, the Contracting Party in whose territory the nuclear installation of that operator is situated shall take such legislative measures as are necessary to enable both that Contracting Party and the other Contracting Parties to benefit from this recourse to the extent that public funds have been made available pursuant to Article 3(b)(ii) and (iii) and (f).

b) Such legislation may provide for the recovery of public funds made available pursuant to Article 3(b) and (iii), and (f) from such operator if the damage results from a fault on his part.

Article 6

In calculating the public funds to be made available pursuant to this Convention, account shall be taken only of those rights to compensation exercised within ten years from the date of the nuclear incident. In the case of damage caused by a nuclear incident involving nuclear fuel or radioactive products or waste which at the time of the incident have been stolen, lost or abandoned and have not yet been recovered, the period shall be ten years from the date of the theft, loss or abandonment. It shall also be extended in the cases and under the conditions laid down in Article 8(b) of the Paris Convention Amendments made to claims after the expiry of this period, under the conditions laid down in Article 8(d) of the Paris Convention, shall also be taken into account.

Article 7

Where a Contracting Party makes use of the right provided for in Article 8(a) of the Paris Convention, the period which it established shall be three years either from the date at which the person suffering damage has knowledge or from the date at which he ought reasonable to have known of both the damage and the operator liable.

Article 8

Any person who is entitled to benefit from the provisions of this Convention shall have the right to full compensation in accordance with national law for damage suffered, provided that, where the amount of damage exceeds or is likely to exceed:

(i) 120 million units of account; or

(ii) If there is aggregate liability under Article 5

Of the Paris Convention and a higher sum results therefrom, such higher sum, any Contracting Party may establish equitable criteria for apportionment. Such criteria shall be applied whatever the origin of the funds and, subjects to the provisions of Article 2, without discrimination based on the nationality, domicile or residence of the person suffering the damage.

Article 9

a) The system of disbursements by which the public funds required under Article (b)(ii) and (iii) and (f) are to be made available shall be that of the Contracting Party whose courts have jurisdiction.

b) Each Contracting Party shall ensure that persons suffering damage may enforce their rights to compensation without having to bring separate proceedings according to the origin of the funds provided for such compensation.

c) No Contracting Party shall be required to make available the public funds referred to in Article 3(b)(ii) and (iii) so long as any of the funds referred to in Article 3(b)(i) remain available.

Article 10

a) The Contracting Party whose courts have jurisdiction shall be required to inform the other Contracting Parties of a nuclear incident and its circumstances as soon as it appears that the damage caused by such incident exceeds, or is likely to exceed, 70 million units of account. The Contracting Parties shall without delay make all the necessary arrangements to settle the procedure for their relations in this connection.

b) Only the Contracting Party whose courts have jurisdiction shall be entitled to request the other Contracting Parties to make available the public funds required under Article 3(b)(iii) and (f) and shall have exclusive competence to disburse such funds.

c) Such Contracting Party shall, when the occasion arises, exercise the right of recourse provided for in Article 5 on behalf of the other Contracting Parties who have made available public funds pursuant to Article 3(b)(iii) and (f).

d) Settlements effected in respect of the payment of compensation out of the public funds referred to in Article 3(b)(ii) and (iii) in accordance with the conditions established by national legislation shall be recognized by the other Contracting Parties, and judgements entered by the competent courts in respect of such compensation shall become enforceable in the territory of the other Contracting Parties in accordance with the provisions of Article 13(e) of the Paris Convention.

Article 11

a) If the courts having jurisdiction are those of a Contracting Party other than the Contracting Party in whose territory the nuclear installation of the operator liable is suitable, the public funds required under Article 3(b)(ii) and (f) shall be made available by the first-named Contracting Party. The Contracting Party in whose territory the nuclear installation of the operator liable is situated shall reimburse to the other Contracting Party the sums paid. These two Contracting Parties shall agree on the procedure of reimbursement.

b) In adopting all legislative, regulatory or administrative provisions, after the nuclear incident occurred, concerning the nature, form and extend of the compensation, the procedure for making available the public funds required under Article 3(b)(ii) and, if necessary, the criteria for the apportionment of such funds, the Contracting Party whose courts have jurisdiction shall consult the Contracting Party in whose territory the nuclear installation of the operator liable is situated. It shall further take all measures necessary to enable the latter to intervene in proceedings and to participate in any settlement concerning compensation.

Article 12

a) The formula for contributions according to which the Contracting Parties shall make available the public funds referred to in Article 3(b)(iii) shall be determined as follows:

(i) as to 50% on the basis of the ratio between the gross national product at current prices of each Contracting Party and the total of the gross national product at current prices of all Contracting Parties as shown by the official statistics published by the Organisation for Economic Co-operation and Development for the year preceding the year in which the nuclear incident occurs;

(ii) as to 50% on the basis of the ratio between the thermal power of the reactors situated in the territory of each Contracting Party and the total thermal power of the reactors situated in the territories of all the Contracting Parties. This calculation shall be made on the basis of the thermal power of the reactors shown at the date of the nuclear incident in the list referred to in Article 2(a)(i); provided that a reactor shall only be taken into consideration for the purposes of this calculation as from the date when it first reaches criticality.

b) For purpose of this Convention, "thermal power" means:

(i) before the issue of a final operating licence, the planned thermal power,

(iii) after the issue of such licence, the thermal power authorized by the competent national authorities.

Article, 13

a) Each Contracting Party shall ensure that all nuclear installations used for peaceful purposes situated in its, territory and falling within the definition in Article 1 of the Paris Convention, appear on the list referred to in Article 2(a)(i).

b) For this purpose, each Signatory or acceding Government shall, on the deposit of its instrument of ratification or accession communicate to the Belgian Government full particulars of such installations.

c) Such particulars shall indicate:

(i) in the case of all installations not yet completed, the expected date on which the risk of a nuclear incident will exist;

(ii) and further, in the case of reactors, the expected date on which they will first reach criticality, and also their thermal power.

d) Each Contracting Party shall also communicate to the Belgian Government the exact date of the existence of the risk of a nuclear incident and, in the case of reactors, the date on which they reached criticality.

e) Each Contracting Party shall also communicate to the Belgian Government all modifications to be made to the list. Where such modifications include the addition of a nuclear installation, the communication must be made at least three months before the expected date on which the risk of nuclear incident will exist.

f) If a Contracting Party is of the opinion that the particulars, or any modification to be made to the list, communicated by another Contracting Party do, not comply with the provisions of Article, 2(a)(i) and of this Article, it may raise objections there to only by addressing them to the Belgian Government within three months from the date on which it has received notice pursuant to paragraph (h) of this Article.

g) If a Contracting Party is of the opinion that a communication required in accordance with this Article has not been made within the time prescribed in this Article, it may raise objections only by addressing them to the Belgian Government within three months from the date on which it knew of the facts which, in its opinion, ought to have been communicated.

h) The Belgian Government shall give notice as soon as possible to each Contracting Party of the communications and objections which it has received pursuant to this Article.

i) The list referred to in Article 2(a)(i) shall consist of all the particulars and modifications referred to in paragraphs (b), (c), (d) and (e) of this Article, it being understood that objections submitted pursuant to paragraphs (f) and (g) of this Article shall have effect retrospective to the date on which they were raised, if they are sustained.

j) The Belgian Government shall supply any contracting party on demand with an up-to-date statement of the nuclear installations covered by this Convention and the details supplied in respect of them pursuant to this Article.

Article 14

a) Except in so far as this Convention otherwise provides, each Contracting Party may exercise the powers vested in it by virtue of the Paris Convention and any provisions made thereunder may be invoked against the other Contracting Parties in order that the public funds referred to in Article 3(b)(ii) and (iii) made available.

b) Any such provisions made by a Contracting Party pursuant to Articles 2, 7(c) and 9 of the Paris Convention as a result of which the public funds referred to in Article 3(b)(ii) and (iii) are required to be made available may not be invoked against any other Contracting Party unless it has consented thereto.

c) Nothing in this Convention shall prevent a Contracting Party from making provisions outside the scope of the Paris Convention and of this Convention, provided that such provided that such provisions shall not involve any further obligation on the part of the other Contracting Parties in so far as their public funds are concerned.

Article 15

a) Any Contracting Party may conclude an agreement with a State which is not a Party to this Convention concerning compensation out of public funds for damage caused by a nuclear incident.

b) To the extent that the conditions for payment of compensation under any such agreement are not more favourable than those which result from measures adopted by the Contracting Party concerned for the application of the Paris Convention and of this Convention, the amount of damage caused by a nuclear incident covered by this Convention and for which compensation is payable by virtue of such an agreement may be taken into consideration, where the proviso to Article 8 applies, in calculating the total amount of damage caused by that incident.

c) The provisions of paragraphs (a) and (b) of this Article shall in no case affect the obligations under Article (b)(ii) and (iii) of those Contracting Parties which have not given their consent to such agreement.

d) Any Contracting Party intending to conclude such an agreement shall notify the other Contracting Parties of its intention. Agreements concluded shall be notified to the Belgian Government.

Article 16

a) The Contracting Parties shall consult each other upon all problems of common interest raised by the application of this Convention and of the Paris Convention, especially Articles 20 and 22 of the later Convention.

b) They shall consult each other on the desirability of revising this Convention after a period of five years from the date of its coming into force and at any other time upon the request of a Contracting Party.

Article 17

Any dispute arising between two or more Contracting Parties concerning the- interpretation or application of this Convention shall, upon the request of a Contracting Party concerned, be submitted to the European Nuclear Energy Tribunal established by the Convention of 20 December 1957 on the Establishment of Security Control in the Field of Nuclear Energy.

Article 18

a) Reservations to one or more of the provisions of this Convention may be made at any time prior to ratification of this Convention if the terms of these reservations have been expressly accepted by all Signatories or at the time of accession or of the application of the provisions of Articles 21 and 24, if the terms of these reservations have been expressly accepted by all Signatories and acceding Governments.

b) Such acceptance shall not be required from a Signatory which has not itself ratified this Convention within a period of twelve months after the date of notification to it of such reservation by the Belgian Government in accordance with Article 25.

c) Any reservation accepted in accordance with the provisions of paragraph (a) of this Article may be withdrawn at any time by notification addressed to the Belgian Government.

Article 19

No State may become or continue to be a Contracting Party to this Convention unless it is a Contracting Party to the Paris Convention.

Article 20

a) The Annex to this Convention shall form an integral part thereof.

b) This Convention shall be ratified. Instruments of ratification shall be deposited with the Belgian Government.

c) This Convention Shall come into force three months after the deposit of the sixth instrument of ratification.

d) For each Signatory ratifying this Convention after the deposit of the sixth instrument of ratification, it shall come into force three months after the date of the deposit of its instrument of ratification.

Article 21

Amendments to this Convention shall be adopted by agreement among all the Contracting Parties. They shall come into force on the date when all Contracting Parties have ratified or confirmed them.

Article 22

a) After the coming into force of this Convention, any Contracting Party to the Paris Convention which has not signed this Convention by notification addressed to the Belgian Government.

b) Such accession shall require the unanimous assent of the Contracting Parties.

c) Once such assent has been given, the Contracting Party to the Paris Convention requesting accession shall deposit its instrument of accession with the Belgian Government.

d) The accession shall take effect three months from the date of deposit of the instrument of accession.

Article 23

a) This Convention shall remain in force until the expiry of the Paris Convention.

b) Any Contracting Party may, by giving twelve months' notice to the Belgian Government, terminate the application of this Convention to itself after the end of the period of ten years specified in Article 22 (a) of the Paris Convention. Within six months after receipt of such notice, any other Contracting Party may, by notice to the Belgian Government, terminate the application of this Convention to itself as from the date when it ceases to have effect in respect of the Contracting Party which first gave notice.

c) The expiry of this Convention or the withdrawal of Contracting Party shall not terminate the obligations assumed by each Contracting Party under this Convention to pay compensation for damage caused by nuclear incidents occurring before the date of such expiry or withdrawal.

d) The Contracting Parties shall, in good time, consult each other on what measures should be taken after the expiry of this Convention or the withdrawal of one or more of the Contracting Parties, to provide compensation comparable to that accorded by this Convention for damage caused by nuclear incidents occurring after the date of such expiry of withdrawal and for which the operator of a nuclear installation in operation before such date within the territories of the Contracting Parties is liable.

Article 24

a) This Convention shall apply to the metropolitan territories of the Contracting Parties.

b) Any Contracting Party desiring the application of this Convention to one or more of the territories in respect of which, pursuant to Article 23 of the Paris Convention, it has given notification of application of that Convention, shall address a request to the Belgian Government.

c) The application of this Convention to any such territory shall require the unanimous assent of the Contracting Parties.

d) Once such asset has been given, the Contracting Party concerned shall address to the Belgian Government a notification which shall take effect as from the date of its receipt.

e) Such notification may, as regards any territory mentioned therein, be withdrawn by the Contracting Party which has made it by giving twelve months notice to that effect to the Belgian Government.

f) If the Paris Convention ceases to apply to any such territory, this Convention shall also cease to apply thereto.

Article 25

The Belgian Government shall notify all Signatories and acceding Governments of the receipt of any instrument of ratification, accession or withdrawal, and shall also notify them of the date on which this Convention comes in to force, the text of any amendment thereto and the date on which such amendment comes into force any reservation made in accordance with Article 18, and all notifications which it has received.

In witness whereof the undersigned Plenipotentiaries, duly empowered, have signed this Convention.

Done at Brussels, this 31st day of January 1963, in the English, Dutch, French, German, Italian and Spanish Languages, the six texts being equally authoritative, in a single copy which shall be deposited with the Belgian Government by whom certified copies shall be communicated to all the other Signatories and acceding Governments.

Annex

To the Convention Supplementary to the Paris Convention of 29 July 1960 on Third Party Liability in the Field of Nuclear Energy

The Governments of the Contracting Parties declare that

compensation for damage caused by a nuclear incident not covered by the Supplementary Convention solely by reason of the fact that the relevant nuclear installation, on account of its utilization, is not on the list referred to in Article 2 of the Supplementary Convention, (including the case where such installation considered by one or more but not all of the Governments to be outside the Paris Convention):

— shall be provided without discrimination among the nationals of the Contracting Parties to the Supplementary Convention; and

— shall not be limited to less than 120 million units of account.

In addition, if they have not already done so, they shall endeavour to make the rules for compensation of persons suffering damage caused by such incidents as similar as possible to those established in respect of nuclear incidents occurring in connection with nuclear installations covered by the Supplementary Convention.

12 Vienna Convention on Civil Liability for Nuclear Damage (1963)

Vienna, 21 May 1963

The Contracting Parties,

Having recognized the desirability of establishing some minimum standards to provide financial protection against damage resulting from certain peaceful uses of nuclear energy,

Believing that a convention on civil liability for nuclear damage would also contribute to the development of friendly relations among nations, irrespective of their differing constitutional and social, systems,

Have decided to conclude a convention for such purposes, and thereto have agreed as follows:

Article I

1. For the purpose of this Convention—

a) "Person" means any individual, partnership, any private or public body whether corporate or not, any international organization enjoying legal personality under the law of the Installation State, and any State or any of its constituent subdivisions.

b) "National of a Contracting Party" includes a Contracting Party or any of its constituent subdivisions, a partnership, or any private or public body whether corporate or not established within the territory of a Contracting Party.

c) "Operator", in relation to a nuclear installation, means the person designated or recognized by the Installation State as the operator of that installation.

d) "Installation State", in relation to a nuclear installation, means the Contracting Party within whose territory that installation is situated or, if it is not situated within the territory of any State, the Contracting Party by which or under the authority of which the nuclear installation is operated.

e) "Law of the competent courts" means the law of the court having jurisdiction under this Convention, including any rules of such law relating to conflict of laws.

f) "Nuclear fuel" means any material which is capable of producing energy by a self-sustaining chain process of nuclear fission.

g) "Radioactive products or waste" means any radioactive material produced in, or any material made radioactive by exposure to the radiation incidental to, the production or utilization of nuclear fuel, but does not include radioisotopes which have reached the final stage of fabrication so as to be usable for any scientific, medical, agricultural, commercial or industrial purpose.

h) "Nuclear material" means—

   (i) nuclear fuel, other than natural uranium and depleted uranium, capable of producing energy by a self-sustaining chair process of nuclear fission outside a nuclear reactor, either alone or in combination with some other material; and

(ii) radioactive products or waste,

i) "Nuclear reactor" means, any structure containing nuclear fuel in such an arrangement that a self-sustaining chain process of nuclear fission can occur therein without an additional source of neutrons.

j) "Nuclear installation" means—

(i) any nuclear reactor other than one with which a means of sea or air transport is equipped for use as a source of power, whether for propulsion thereof or for any other purpose;

(ii) any factory using nuclear fuel for the production of nuclear material, or any factory for the processing of nuclear material, including any factory for the re-processing of irradiated nuclear fuel; and

(iii) any facility where nuclear material is stored, other than storage incidental to the carriage of such material;

provided that the Installation State may determine that several nuclear installations of one operator which are located at the same site shall be considered as a single nuclear installation.

k) "Nuclear damage" means—

(i) loss of life, and personal injury or any loss of, or damage to, property which arises out of or results from the radioactive properties or a combination of radioactive properties with toxic, explosive or other hazardous properties of nuclear fuel or radioactive products or waste in, or of nuclear material coming from, originating in, or sent to a nuclear installation.

(ii) any other loss of damage so arising or resulting if and to the extent that the law of the competent court so provides; and

(iii) if the law of the Installation State so provides, loss of life, any personal injury or any loss of, or damage to, property which arises out of or results from other ionizing radiation emitted by any other source of radiation inside a nuclear installation.

1) "Nuclear incident" means by any occurrence or series of occurrences having the same origin which causes nuclear damage.

2. An Installation State may, if the small extent of the risks involved so warrants, exclude any small quantities of a nuclear material from the application of this Convention, provided that—

a) maximum limits for the exclusion of such quantities have been established by the Board of Governors of the International Atomic Energy Agency; and

b) any exclusion by an Installation State is within such established limits.

The maximum limits shall be reviewed periodically by the Board of Governors.

Article II

1. The operator of a nuclear installation shall be liable for nuclear damage upon proof that such damage has been caused by a nuclear incident—

a) in his nuclear installation; or

b) involving nuclear material from or originating in his nuclear installation, and occurring—

(i) before liability with regard to nuclear incidents involving the nuclear material has been assumed, pursuant to the express terms of a contract in writing, by the operator of another nuclear installation;

(ii) in the absence of such express terms, before the operator of another nuclear installation has taken charge of the nuclear material; or

(iii) where the nuclear material is intended to be used in nuclear reactor with which a means of transport is equipped for use as a source of power, whether for propulsion thereof or for any other purpose, before the person duly authorized to operate such reactor has taken charge of the nuclear material; but

(iv) where the nuclear material has been sent to a person within the territory of a non-Contracting State, before it has been unloaded from the means of transport by which it has arrived in the territory of that non-Contracting State;

c) involving nuclear materials to his nuclear installation, and occurring—

(i) after liability with regard to nuclear incidents involving the nuclear material has been assumed by him, pursuant to the express terms of a contract in writing, from, the operator of another nuclear installation;

(ii) in the absence of such express terms, after he has taken charge of the nuclear material; or

(iii) after he has taken charge of the nuclear material from a person operating a nuclear reactor with which a means of transport is equipped for use as a source of power, whether for propulsion thereof or for any other purpose; but

(iv) where the nuclear material has, with the written consent of the operator, been sent from a person within the territory of a Contracting State, only after it has been loaded on the means of transport by which it is to be carried from the territory of that State;

provided that, if nuclear damage is caused by a nuclear incident occurring in a nuclear installation and involving nuclear material stored therein incidentally to the carriage of such material, the provisions of sub-paragraph (a) of this paragraph shall not apply where another operator or person is solely liable pursuant to the provisions of sub-paragraph (b) or (c) of this paragraph.

2. The Installation State may provide by legislation that, in accordance with such terms as may be specified therein, a carrier of nuclear material or a person handling radioactive waste may, at his request and with the consent of the operator concerned, be designated or recognized as operator in the place of that operator in respect of such nuclear material or radioactive waste respectively. In this case such carrier or such person shall be considered, for all the purposes of this Convention, as an operator of a nuclear installation situated within the territory of that State.

3. a) Where nuclear damage engages the liability of more than one operator, the operators involved shall, in so far as the damage attributable to each operator is not reasonably separable, be jointly and severally liable.

b) Where a nuclear incident occurs in the course of carriage of nuclear material, either in one and the same means of transport, or, in the case of storage incidental to the carriage, in one and the same nuclear installation, and causes nuclear damage which engages the liability of more than one operator, the total liability shall not exceed the highest amount applicable with respect to any one of them pursuant to Article V.

c) In neither of the cases referred to in sub-paragraphs (a) and (b) of this paragraph shall the liability of any one operator exceed the amount applicable with respect to him pursuant to Article V.

4. Subject to the provisions of paragraph 3 of this Article, where several nuclear installations of one and the same operator are involved in one nuclear incident, such operator shall be liable in respect of each nuclear installation involved up to the amount applicable with respect to him pursuant to Article V.
5. Except as otherwise provided in this Convention, no person other than the operator shall be liable for nuclear damage. This, however, shall not affect the application of any international convention in the field of transport in force or open for signature, ratification or accession at the date on which this Convention is opened for signature.
6. No person shall be liable for any loss or damage which is not nuclear damage pursuant to sub-paragraph (k) of paragraph 1 of Article I but which could have been included as such pursuant to sub-paragraph (k)(ii) of that paragraph.
7. Direct action shall lie against the person furnishing financial security pursuant to Article VII, if the law of the competent court so provides.

Article III

The operator liable in accordance with this Convention shall provide the carrier with a certificate issued by or on behalf of the insurer or other financial guarantor furnishing the financial security required pursuant to Article VII. The certificate shall state the name and address of that operator, and the amount, type and duration of the security, and these statements may not be disputed by the person by whom or on whose behalf the certificate was issued. The certificate shall also indicate the nuclear material in respect of which the security applies and shall include a statement by the competent public authority of the Installation State that the person named is an operator within the meaning of this Convention.

Article IV

1. The liability of the operator for nuclear damage under this Convention shall be absolute.

2. If the operator proves that the nuclear damage resulted wholly or partly either from the gross negligence of the person suffering the damage or from an act or omission of such person done with intent to cause damage, the competent court may, if its law so provides, relieve the operator wholly or partly from his obligation to pay compensation in respect of the damage suffered by such person.

3. a) No liability under this Convention shall attach to an operator for nuclear damage caused by a nuclear incident directly due to an act of armed conflict, hostilities, civil war or insurrection.

b) Except in so far as the law of the Installation State may provide to the contrary, the operator shall not be liable for nuclear damage caused by a nuclear incident directly due to a grave natural disaster of an exceptional character.

4. Whenever both nuclear damage and damage other than nuclear damage have been caused by a nuclear incident or jointly by a nuclear incident and one or more other occurrences, such other damage shall, to the extent that it is not reasonably separable from the nuclear damage, be deemed, for the purposes of this Convention, to be nuclear damage caused by that nuclear incident. Where, however, damage is caused jointly by a nuclear incident covered by this Convention and by an emission of ionizing radiation not covered by it nothing in this Convention shall

limit or otherwise affect the liability, either as regards any person suffering nuclear damage or by way of recourse or contribution, of any person who may be held liable in connection with that emission of ionizing radiation.

5. The operator shall not be liable under this Convention for nuclear damage—

   a) to the nuclear installation itself or to any property on the site of that installation which is used or to be used in connection with that installation; or

   b) to the means of transport upon which the nuclear material involved was at the time of the nuclear incident.

6. Any Installation State may provide by legislation that sub-paragraph (b) of paragraph 5 of this Article shall not apply, provided that in no case shall the liability of the operator in respect of nuclear damage, other than nuclear damage to the means of transport, be reduced to less than US$5 million for any one nuclear incident.

7. Nothing in this Convention shall affect—

   a) the liability of any individual for nuclear damage for which the operator, by virtue of paragraph 3 or 5 of this, Article, is not liable under this Convention and which that individual caused by an act or omission done with intent to cause damage; or

   b) the liability outside this Convention of the operator for nuclear damage for which, by virtue of sub-paragraph (b) of paragraph 5 of this Article, he is not liable under this Convention.

Article V

1. The liability of the operator may be limited by the Installation State to not less than US$5 million for any one nuclear incident.
2. Any limits of liability which may be established pursuant to this Article shall not include any interest or costs awarded by a court in actions for compensation of nuclear damage.
3. The United States dollar referred to in this Convention is a unit of account equivalent to the value of the United States dollar in terms of gold on 29 April 1963, that is to say US$35 per one troy ounce of fine gold.
4. The sum mentioned in paragraph 6 of Article IV and in paragraph 1 of this Article may be converted into national currency in round figures.

Article VI

1. Rights of compensation under this Convention shall be extinguished if an action is not brought within ten years from the date of the nuclear incident. If, however, under the law of the Installation State the liability of the operator is covered by insurance or other financial security or by State funds for a period longer than ten years, the law of the competent court may provide that rights of compensation against the operator shall only be extinguished after a period which may be longer than ten years, but shall not be longer than the period for which his liability is so covered under the law of the Installation State. Such extension of the extinction period shall in no case affect rights of compensation under this Convention of any person who has brought an action for loss of life or personal injury against the operator before the expiry of the aforesaid period of ten years.

2. Where nuclear damage is caused by a nuclear incident involving nuclear material which at the time of the nuclear incident was stolen, lost, jettisoned or abandoned, the period established pursuant to paragraph 1 of this Article shall be computed from the date of that nuclear incident, but the period shall in no case exceed a period of twenty years from the date of the theft, loss, jettison or abandonment.

3. The law of the competent court may establish a period of extinction or prescription of not less than three years from the date on which the person suffering nuclear damage had knowledge or should have had knowledge of the damage and of the operator liable for the damage provided that the period established pursuant to paragraphs 1 and 2 of this Article shall not be exceeded.

4. Unless the law of the competent court otherwise provides, any person who claims to have suffered nuclear damage and who has brought an action for compensation within the period applicable pursuant to this Article may amend his claim to take into account any aggravation of the damage, even after the expiry of that period, provided that final judgment has not been entered.

5. Where jurisdiction is to be determined pursuant to sub-paragraph (b) of paragraph 3 of Article XI and a request has been made within the period applicable pursuant to this Article to any one of the Contracting Parties empowered so to determine, but the time remaining after such determination is less than six months, the period within which an action may be brought shall be six months, reckoned from the date of such determination.

Article VII

1. The operator shall be required to maintain insurance or other financial security covering his liability for nuclear damage in such amount, of such type and in such terms as the Installation State shall specify. The Installation State shall ensure the payment of claims for compensation for nuclear damage which have been established against the operator by providing the necessary funds to the extent that the yield of insurance or other financial security is inadequate to satisfy such claims, but not in excess of the limit, if any, established pursuant to Article V.
2. Nothing in paragraph 1 of this Article shall require a Contracting Party or any of its constituent subdivisions, such as States or Republics, to maintain insurance or other financial security to cover their liability as operators.
3. The funds provided by insurance, by other financial security or by the Installation State pursuant to paragraph 1 of this Article shall be exclusively available for compensation due under this Convention.
4. No insurer or other financial guarantor shall suspend or cancel the insurance or other financial security provided pursuant to paragraph 1 of this Article without giving notice in writing of at least two months to the competent public authority or, in so far as such insurance or other financial security relates to the carriage of nuclear material, during the period of the carriage in question.

Article VIII

Subject to the provision of this Convention, the nature, form and extent of the compensation, as well as the equitable distribution thereof, shall be governed by the law of the competent court.

Article IX

1. Where provisions of national or public health insurance, social insurance, social security, workmen's compensation or occupational disease compensation systems include compensation for nuclear damage, rights beneficiaries of such system to obtain compensation under this Convention and rights of recourse by virtue of such systems against the operator liable shall be determined, subject to the provisions of this Convention, by the law of the Contracting Party in which such systems have been established, or by the regulations of the intergovernmental organization which established such systems.

2. a) If a person who is a national of a Contracting Party, other than the operator, has paid compensation for nuclear damage under an international convention or under the law of a non-Contracting State, such person shall, up to the amount which he has paid, acquire by subrogation the rights under this Convention of the person so compensated. No rights shall be so acquired, by any person to the extent that the operator has a right of recourse against such person under this Convention.

b) Nothing in this Convention shall preclude an operator who has paid compensation for nuclear damage out of funds other than those provided pursuant to paragraph to paragraph 1 of Article VII from recovering from the person providing financial security pursuant to that paragraph or from the Installation State, up to the amount he has paid, the sum which the person so compensated would have obtained under this Convention.

Article X

The operator shall have a right of recourse only—

a) if this is expressly provided for by a contract in writing; or

b) if the nuclear incident results from an act or omission done with intent to cause damage, against the individual who has acted or omitted to act with such intent.

Article XI

1. Except as otherwise provided in this Article, jurisdiction over action under Article II shall lie only with the courts of the Contracting Party within whose territory the nuclear incident occurred.

2. Where the nuclear incident occurred outside the territory of any Contracting Party, or where the place of the nuclear incident cannot be determined with certainty, jurisdiction over such actions shall lie with the courts of the Installation State of the operation liable.

3. Where under paragraph 1 or 2 of this Article, jurisdiction would lie with the courts of more than one Contracting Party, jurisdiction shall lie—

   a) if the nuclear incident occurred partly outside the territory of any Contracting Party, and partly within the territory of a single Contracting Party, with the courts of the latter; and

   b) in any other case, with the courts of that Contracting Party which is determined by agreement between the Contracting Parties whose courts would be competent under paragraph 1 or 2 of this Article.

Article XII

1. A final judgment entered by a court having jurisdiction under Article XI shall be recognized within the territory of any other Contracting Party, except—
   a) where the judgment was obtained by fraud;
   b) where the party against whom the judgment was pronounced was not given a fair opportunity to present his case; or
   c) where the judgment is contrary to the public policy of the Contracting Party within the territory of which recognition is sought, or is not accord with fundamental standards of justice.
2. A final judgment which is recognized shall, upon being presented for enforcement in accordance with the formalities required by the law of the Contracting Party where enforcement is sought, be enforceable as if it were a judgment of a court of that Contracting Party.
3. The merits of a claim on which the judgment has been given shall not be subject to further proceedings.

Article XIII

This Convention and the national law applicable thereunder shall be applied without any discrimination based upon nationally, domicile or residence.

Article XIV

Except in respect of measures of execution, jurisdiction immunities under rule of national or international law shall not be invoked in actions under this Convention before the courts competent pursuant to Article XI.

Article XV

The contracting Parties shall take appropriate measures to ensure that compensation for nuclear damage, interest and costa awarded by a court in connection therewith, insurance and reinsurance premiums and funds provided by insurance, reinsurance or other financial security, or funds provided by the Installation State, pursuant, to this Convention shall be freely transferable into the currency of the Contracting Party within whose territory the damage is suffered, and of the Contrasting Party within whose territory the claimant is habitually resident, and as regards insurance or reinsurance premiums and payments, into the currencies specified in the insurance or reinsurance contract.

Article XVI

No person shall be entitled to recover compensation under this Convention to the extent that he has recovered compensation in respect of the same nuclear damage under another international convention on civil liability in the field of nuclear energy.

Article XVII

This Convention shall not, as between the parties to them affect the application of any international agreements or international convention on civil liability in the field of nuclear energy in force, or open for signature, ratification or accession at the date on which this Convention is opened for signature.

Article XVIII

This Convention shall not be construed as affecting the rights, if any of a Contracting Party under the general rules of public international law in respect of nuclear damage.

Article XIX

1. Any Contracting Party entering into an agreement pursuant to sub-paragraph (b) of paragraph 3 of Article XI shall furnish without delay to the Director General of the International Atomic Energy Agency for information and dissemination to the other Contracting Parties a copy of such agreement.
2. The Contracting Parties shall furnish to the Director-General for information and dissemination to the other Contracting Parties copies of their respective laws and regulations relating to matters covered by this Convention.

Article XX

Notwithstanding the termination of the application of this Convention to any Contracting Party, either by termination pursuant to Article XXV or by denunciation pursuant to Article XXVI, the provisions of this Convention shall continue to apply to any nuclear damage caused by a nuclear incident occurring before such termination.

Article XXI

This Convention shall be open for signature by the States represented at the International Conference on Civil Liability for Nuclear Damage held in Vienna form 29 April to 19 May 1963.

Article XXII

This Convention shall be ratified, and the instruments of ratification shall be deposited with the Director General of the International Atomic Energy Agency.

Article XXIII

This Convention shall come into force three months after the deposit of the fifth instrument of ratification, and in respect of each State ratifying it thereafter, three months after the deposit or the instrument of ratification by that State.

Article XXIV

1. All States Members of the United Nations, or of any of the Specialized agencies or of the International Atomic Energy Agency not represented at the International Conference on Civil Liability for Nuclear Damage, held in Vienna from 29 April to 19 May 1963, may accede to this Convention.
2. The instruments of accession shall be deposited with the Director General of the International Atomic Energy Agency.
3. This Convention shall come into force in respect of the acceding State three months after the date of deposit of the instrument of accession of that State but not before the date of the energy into force of this Convention pursuant to Article XXIII.

Article XXV

1. This Convention shall remain in force for a period of ten years from the date of its entry into force. Any Contracting Party may, by giving before the end of that period at least twelve months notice to that effect to the Director General of the International Atomic Energy Agency, terminate the application of this Convention to itself at the end of that period of ten years.
2. This Convention shall, after that period of ten years, remain in force for a further period of five years for such Contracting Parties as have not terminated its application pursuant to paragraph 1 of this Article, and thereafter for successive periods of five years each for those Contracting Parties which have not terminated its application at the end of one of such periods, by giving, before the end of one of such

periods, at lease twelve months notice to that effect to the Director General of the International Atomic Energy Agency.

Article XXVI

1. A conference shall be convened by the Director General of the International Atomic Energy at any time after the expiry of period of five years from the date of the entry into force of this Convention in order to consider the revision thereof, if one-third of the Contracting Parties express a desire to that effect.
2. Any Contracting Party may denounce this Convention by notification to the Director General of the International Atomic Energy Agency within a period of twelve months following the first revision conference, held pursuant to paragraph 1 of this Article.
3. Denunciation shall take effect one year after the date on which notification to that effect has been received by the Director General of the International Atomic Energy Agency.

Article XXVII

The Director General of the International Atomic Energy Agency shall notify the States invited to the International Conference on Civil liability for Nuclear Damage held in Vienna from 29 April to 19 May 1963 and the States which have acceded to this Convention of the following:

a) signatures and instruments of ratification and accession received pursuant to Articles XXI, XXII, and XXVII;

b) the date on which this Convention will come into force pursuant to Article XXVIII;

c) notifications of termination and denunciation received pursuant to Articles XXV and XXVI;

d) requests for the convening of a revision conference pursuant to Article XXVI.

Article XXVIII

This Convention shall be registered by the Director General of the International Atomic Energy in accordance with Article 102 of the Charter of the United Nations.

Article XXIX

The original of this Convention, of which the English, French, Russian and Spanish texts are equally authentic, shall be deposited with the Director General of the International Atomic Energy Agency, who shad issue certified copies.

In Witness Whereof, the undersigned Plenipotentiaries, duly authorized thereto, have signed this Convention.

Done in Vienna, this twenty-first day of May, one thousand nine hundred and sixty-three.

INDEX